T0206331

Lecture Notes in Energy

Volume 81

Lecture Notes in Energy (LNE) is a series that reports on new developments in the study of energy: from science and engineering to the analysis of energy policy. The series' scope includes but is not limited to, renewable and green energy, nuclear, fossil fuels and carbon capture, energy systems, energy storage and harvesting, batteries and fuel cells, power systems, energy efficiency, energy in buildings, energy policy, as well as energy-related topics in economics, management and transportation. Books published in LNE are original and timely and bridge between advanced textbooks and the forefront of research. Readers of LNE include postgraduate students and non-specialist researchers wishing to gain an accessible introduction to a field of research as well as professionals and researchers with a need for an up-to-date reference book on a well-defined topic. The series publishes single- and multi-authored volumes as well as advanced textbooks.

Indexed in Scopus and EI Compendex The Springer Energy board welcomes your book proposal. Please get in touch with the series via Anthony Doyle, Executive Editor, Springer (anthony.doyle@springer.com)

More information about this series at http://www.springer.com/series/8874

Alice J. Friedemann

Life After Fossil Fuels

A Reality Check on Alternative Energy

 Springer

Alice J. Friedemann
Oakland, CA, USA

ISSN 2195-1284   ISSN 2195-1292   (electronic)
Lecture Notes in Energy
ISBN 978-3-030-70337-0   ISBN 978-3-030-70335-6   (eBook)
https://doi.org/10.1007/978-3-030-70335-6

This Springer imprint is published by the registered company Springer Nature Switzerland AG
The registered company address is: Gewerbestrasse 11, 6330 Cham, Switzerland

Preface

Fellow shareholders in the future of the planet Earth, I am not going to sugarcoat it. We have trouble ahead. In this book, we are going to look at the outlook for substitute, alternative energy resources come the day that fossil fuels decline. This is not going to be a quarterly report where we look ahead three months. Rather, we are going to be peering through the proper end of the telescope, focusing on the viability of energy alternatives for the years ahead, for the next generations.

Humanity has made inconceivable progress in the past 500 years. Books once were created one at a time by hand. The printing press opened the door to a revolution of knowledge. Cement and steel created the foundations of modern civilization. Wooden ships have been replaced by container ships that cross the oceans bringing fish from Chile, wines from France, automobiles from Japan, and everything from China. In 1869, the first transcontinental railroad crossed America, and 100 years later, the first human stepped foot on the moon.

Imagination and invention are the sparks that illumined this path. We humans can pat ourselves on the back for that. Ah, didn't that feel good! Yet all the while, our progress was fueled and enabled by something not of our own making, fossil fuels. From the Industrial Revolution, fired by coal, to today's global technology society, fossil fuels have been the indispensable element. Economies are about work, and one barrel of oil does the work a strong human can do in four and a half years.

Today, most of humanity realizes that progress and the planet itself is threatened by climate change. Climate change is finally front-page news, and it is said to be the main threat facing humanity. It is and will continue to be catastrophic, affecting us for many centuries, if not millennia to come.

Climate change is an existential threat. Agreed. Yet I believe that we are ignoring an even greater threat, one that is not front and center. This book makes the case that the deadliest crisis facing our civilization is energy decline. Peak oil production may have already occurred. If that is true, then society should be working hard, in fact should be on a war footing, to prepare for life without plentiful, affordable fossil fuels.

Energy usage and climate change are joined at the hip. Many means are being developed to cut carbon emissions including enhancing the energy efficiency of buildings, fleets of electric cars, deploying millions of solar panels and wind turbines, and carbon sequestration. Maybe not in the short run, but ultimately, energy decline will mitigate climate change more than all of these approaches. As oil, coal, and natural gas inevitably decline, carbon emissions will be dramatically reduced.

This book was inspired by my book on transportation, "When Trucks Stop Running, Energy and the Future of Transportation" (Friedemann 2016). Leveraging this research, here I widen my focus and examine our broader dependence on fossil fuels, not just for transportation, but also for manufacturing, agriculture, and industry. The book devotes significant attention to farming and what will be possible absent affordable fossil fuels. My 2017 essay, "Peak Soil" (Friedemann 2017) at my blog www.energyskeptic.com previewed the importance and neglect of farmland soils, a landscape that is more fully explored here.

Though they are ubiquitous in our lives, for the most part, we take fossil fuels for granted. Conventional economics pays little attention to this indispensable component because it is relatively inexpensive—which of course is why it is so valuable. This book remedies that examining the critical sectors of our society that rely on fossil fuels and soberly assesses the scores of alternative energy options under development. Which alternatives have promise? What are the challenges for making them commercial? Which alternatives are mere wishful thinking? After the last page has been turned, you will have in hand a reality check on where our energy will come from in the future and the sustainability of our way of life.

References

Friedemann AJ (2016) When Trucks Stop Running, Energy and the Future of Transportation. Springer.
Friedemann AJ (2017) Peak soil: Industrial agriculture destroys ecosystems and civilizations. Biofuels make it worse. http://energyskeptic.com/2017/peaksoil/. Accessed 18 Nov 2020

Oakland, CA   Alice J. Friedemann

Acknowledgments

I am most grateful to my beloved husband, journalist, and science writer Jeffery B. Kahn, who edited this book through six drafts, making it more accessible, inviting in readers who are not already academic energy experts. He added a light-hearted touch and infusions of humor to leaven the difficult subject this book confronts. As with daily breakfast-table repartee between us, there are two intertwined voices that a careful reader will discern in this book. Together, I hope they both enlighten and entertain you. Jeffery is a superb companion to have in this hermit life of Covid-19, full of joy and helping me find more wonderment in the world.

At peak oil conferences, I had the good fortune to meet professor Charles Hall, who introduced me to the concept of energy returned on energy invested, which he mainly developed, and his must-read book "Energy and the Wealth of Nations: An Introduction to Biophysical economics." Being the principal editor of this book, he received his PhD in Systems Ecology from Howard Odum at the University of North Carolina at Chapel Hill and spent more than half a century doing research and teaching at Cornell University, the University of Montana, and SUNY.

At these peak oil conferences, I also met Professor Albert Bartlett who impressed upon me the importance of exponential growth, and Professor Walter Youngquist who wrote one of the best overviews of energy and resources in his book, "Geodestinies."

Others I am grateful to have met or read include John Perlin ("A Forest Journey"), Richard Heinberg ("Afterburn"), Gail Tverberg (Ourfiniteworld.com), Vaclav Smil ("Prime Movers" and "Harvesting the Biosphere"), Donella Meadows et al. ("Limits to Growth"), Jason Bradford ("The Future is Rural"), Pedro Prieto & Charles Hall ("Spain's Photovoltaic Revolution"), David Pimentel ("Food, Energy, and Society"), David Hughes ("Drilling Deeper"), William Alley ("Too Hot to Touch"), David Fridley ("Our Renewable Future"), Euan Mearns (euanmearns.com), Nate Hagens ("The Bottlenecks of the 21st Century" & YouTube videos), Ugo Bardi ("Extracted"), Garrett Hardin ("Living within Limits"), Alan Weisman ("Countdown" & "The World Without Us"), William Rees ("Our Ecological Footprint"), Rex Weyler (rex-weyler.ca), Marc Levinson ("The Box"), Eleanor Agnew ("Back From The Land"), E.O. Wilson ("The Social Conquest of Earth"), Richard Wrangham ("Catching Fire"), Peter Ward ("Rare Earth"), Kurt Andersen ("Fantasyland"), Naomi Oreskes

("Merchants of Doubt"), Carl Sagan ("The Demon-Haunted World"), Michael Shermer ("The Science of Good & Evil"), and hundreds more as detailed in my book lists at http://energyskeptic.com/category/books/book-list/

Contents

Chapter 1
Introduction

A Canadian petroleum refinery working long into the night. (Photo credit Kurayba 2017)

The Coming Energy Crisis

Even more than most of us realize, we are completely and utterly dependent on fossil fuels. Oil makes everything possible. Cement, steel, roads, cars, farm machinery, food, health care, and 500,000 products. I would really like to list these products for you, perhaps just a sampling of 100,000 or so. Sorry, but my publisher tells me there is not enough ink for that. Okay, just two: Ink and this book!

A. J. Friedemann, *Life after Fossil Fuels*, Lecture Notes in Energy 81, https://doi.org/10.1007/978-3-030-70335-6_1

There was a time before gas stations. The first commercial oil well in the US was drilled in 1859. Since then, we have burned a lot of oil, some 1.3 trillion barrels of it. Today, there are 7.8 billion people on the Earth, most of us living on the back of oil. Together, we human beings burn one cubic mile of oil every year.

This can not go on forever. Let us agree right here at the outset that fossil fuels are finite and one day will need to be replaced with something else. Global oil production will certainly peak and then decline unless the Earth is actually a giant gas tank refilled by a Petroleum God.

The end of cheap and easy oil ended in 2005 when conventional oil plateaued. From now on the cost and difficulty of obtaining oil will increase. Regardless of the exact year that production of fossil fuels peaks, we have known for 70 years that models predicted peak world production around now (Hubbert 1956; Deffeyes 2010; Inman 2016), and the population continues to grow, driving up the need for it.

Although many assumed that a sign of peak oil would be high prices, low prices may actually signal future decline. This is because the fossil industry can not afford to start new projects because of low prices (Tverberg 2018). And in a depression, which may be precipitated by the pandemic, companies can not readily go to their friendly banker and borrow billions of dollars to find and drill for more oil.

The uneven and unfair distribution of wealth is another obstacle to raising the price of oil enough for new drilling projects. If oil prices rise substantially, for many people the price would be unaffordable. This could trigger another down economic cycle. In the US, the top 1% owns 31% of the wealth, the bottom half just 1.4%.

No, geological depletion is not the only reason an energy crisis can arise. An energy crisis can be triggered by drought, heatwaves, wars, social unrest, terrorism, hurricanes, and floods. Oil-producing nations with growing populations may reduce their exports, and those past peak oil production, such as Syria, Yemen, Egypt, and Nigeria, may be unable to afford cheap food and oil for their citizens (Ahmed 2017).

Oil production can decline due to other issues besides depletion. Venezuela still has among the greatest reserves of any country but has not done proper maintenance so that its oil production in 2019 was only 18% of what it was two decades ago. Russia's vast reserves are controlled by a corrupt kleptocracy that may run their oil and gas business right into the ground. Russian kleptocrats are not maintaining existing infrastructure or investing in new oilfields (Stratfor 2020; Gustafson 2012; Maddow 2019).

Time's a Wastin'

In this book, I make the case that energy decline is the greatest existential crisis humanity faces, or has ever faced. As geologist Walter Youngquist once wrote: "Peak oil will affect more people, in more places, in more ways, than anything else in the history of the world."

History shows that it usually takes from 20 to 70 years for a new energy technology to go from its first experimental stage to reach one 1% of a national market. For example, it took 30 years for the first prototype lithium-ion batteries to be commercialized, and 25 years for solar photovoltaic to reach a 1% share of nation's electricity supply in Spain and Germany.

Thus, we are running out of time for maintaining the energy supplies we need. For the foreseeable future, we will still have oil, just less of it. Peak oil does not mean RUNNING OUT OF OIL. It means global oil production has peaked, and there is less to be had after that. Peak oil means that some current uses of oil will no longer be possible. Understand that and you understand that now is the time to redeploy the oil that remains, allocating it to only uses where there are no easy substitutes, and using other fossil fuels to transition to a simpler world.

A replacement for fossils must be commercial *soon* if we are past peak production or close to it. Vast amounts of oil will be required to build new energy systems and supporting infrastructure. Nuclear, photovoltaics, wind, hydro, batteries—name your alternative—it can not be built without oil. But as oil production declines, petroleum will have to be rationed to agriculture and other essential services (DOE 1980).

This book explores what will probably happen after peak oil. What are our options for replacing the fossil fuels that turn the great wheel of civilization? What alternatives should we deploy right now? Which technologies merit further research and development? Which ones are mere wishful thinking that, upon careful scrutiny, dematerialize before our eyes? Spoiler alert: What lies ahead is not easy to face and will not resemble civilization as we know it now.

The first few chapters of this book explore the ways we depend on oil, coal, and natural gas. Some of these dependencies involve the core ingredients of civilization itself yet are largely unknown and will surprise you. Next, the book looks at hydrogen, wind, solar, geothermal, nuclear, batteries, catenary systems, fusion, methane hydrates, power2gas, wave, and tidal power and examine whether they can replace or even supplement fossil fuels' role in transportation, manufacturing, fertilizers, and electricity. And finally, we will examine what liquid biofuels or combustible biomass can do for us.

Will this new energy portfolio be up to the job? As you read along, I invite you to make your own judgments. Ultimately, I conclude that for core civilization services—manufacturing, transportation, and even the electric grid—this arsenal of alternatives will be disappointing. Taking off the rose-colored glasses, I examine the challenges, identify the showstoppers.

The media promote the belief that with renewable electricity we can carry on consuming goods and the endless growth trajectory we have been on since coal fired the Industrial Revolution in the eighteenth century. One group of scientists, Jacobson et al. (2015a, b), contend that solar, wind, and other electric power can completely replace fossils, Save the World and allow us to continue more or less as we are today. I am going to show why Jacobson is wrong. I join with Clack et al. (2017) and will document why their plan is a pipedream.

Ultimately, I am going to make the case that after fossil fuels decline, civilization will change in ways that are almost impossible to contemplate, returning again to biomass, mainly wood, for thermal energy and infrastructure, like all civilizations before coal. Not surprising really, since wood was a source of energy in all civilizations before fossil fuels (Perlin 2005).

We are fated to live in a Wood World. Biomass is the one known energy source that can do the job of several of the critical civilization services performed by oil. In this new Wood World, there will always be too many people, too much energy demand, and never enough biomass. Civilization as we know it now will look like science fiction.

Since the coming energy crisis is not headline news, I document what I write with hundreds of citations of published science. I realize these citations can be distracting to the narrative flow and your light, summer-reading pleasure. But I include them to make transparent the sources and breadth of research involved in this book and to give you an opportunity to explore a topic more deeply. If you do not find what I write to be credible, look up the source I cite for details.

Throughout the book, I will identify actions that can be taken, but this crisis has no solution that can sustain the status quo. For those looking for an action list, the final chapter will be particularly useful.

If we face our Wood World future, we could make life far better for our descendants by using our still plentiful energy to redesign infrastructure, buildings, add insulation, install heat pumps, plant forests, and expand organic agriculture. If we all consumed less energy, we could stretch out the time we have left to make this transition. Most actions we take also will reduce climate change. And make life better right now for us.

Over the past decade, we have been warned that civilization is under threat from a global pandemic. We have been warned about the threat of climate change. Our collective lack of response indicates that we find such threats unthinkable. Overwhelmed, we look away. We have ignored these warnings, and in so doing imperiled the planet, our nations, and our own personal safety.

What I am writing about in this book also seems impossible, particularly in this moment of low oil prices. Yet, just as certain as oil is finite, there is an energy crisis coming in the future.

As the energy crisis unfolds, the blame will be cast on oil companies, the government, the rich, and baby boomers. Conspiracy theories will have a golden age.

But anyone who wants to cast the first stone will have to look to history, to go back to sixteenth century England when deforestation forced Londoners to burn coal. That led to the fossil-fueled Industrial Revolution and a one-time only civilization and population explosion. Fueled by oil, coal, and natural gas, we temporarily escaped the constraints of the carrying capacity of our planet (Cavert 2016). As fossil fuels peak and then decline, we will fall back to Earth.

References

Ahmed N (2017) Failing states, collapsing systems: bio physical triggers of political violence. Springer

Cavert WM (2016) The smoke of London: energy and environment in the early modern city. Cambridge University Press

Clack CTM, Qvist SA, Apt J et al (2017) Evaluation of a proposal for reliable low-cost grid power with 100% wind, water, and solar. Proc Natl Acad Sci

Deffeyes KS (2010) When oil peaked. Hill and Wang

DOE (1980) Standby Gasoline Rationing Plan. DOERG-0029. Dist category UC UC-92. U.S. Department of Energy, Office of Regulations and Emergency Planning

Gustafson T (2012) Wheel of Fortune. The Battle for Oil and Power in Russia. Harvard University Press

Hubbert MK (1956) Nuclear energy and the fossil fuels. Spring Meeting of the Southern District, American Petroleum Institute, Plaza Hotel, San Antonio, Texas, March 7–8–9, 1956

Inman M (2016) The oracle of oil: a maverick geologist's quest for a sustainable future. W.W. Norton & Company

Jacobson MZ, Delucchi MA, Cameron MA et al (2015a) Low-cost solution to the grid reliability problem with 100% penetration of intermittent wind, water, and solar for all purposes. Proc Natl Acad Sci U S A 112:15060–15065

Jacobson MZ, Delucchi MA, Bazouin G et al (2015b) 100% clean and renewable wind, water, and sunlight (WWS) all-sector energy roadmaps for the 50 United States. Energy Environ Sci 8:2093–2117

Kurayba (2017) Image "Radiant Refinery" by Kurayba is licensed with CC BY-SA 2.0. To view a copy of this license. https://creativecommons.org/licenses/by-sa/2.0/

Maddow R (2019) Blowout: corrupted democracy, rogue state Russia, and the richest, most destructive industry on earth. Crown

Perlin J (2005) A forest journey: the story of wood and Civilization. Countryman Press

Stratfor (2020) The golden age of russian oil nears an end. https://worldview.stratfor.com/article/golden-age-russian-oil-nears-end-energy-economy-shale-crude. Accessed 1 Nov 2020

Tverberg G (2018) Low oil prices: an indication of major problems ahead? https://ourfiniteworld.com/2018/11/28/low-oil-prices-an-indication-of-major-problems-ahead/. Accessed 1 Nov 2020

Chapter 2
We Are Running Out of Time

I like to think of the oil age as beginning in 1901 with the exploding geyser of oil at Spindletop sending plumes 150 ft into the Texas air (Fig. 2.1). At Spindletop, oil gushed to the surface from natural pressure, no pumping required. This is what is known as the best of "conventional" oil, essentially manna from down below. In conventional oil fields like this, typically the first 5–15% of the oil gushes up out of the ground of its own accord as if through the ages it had awaited its liberation. For the petroleum business, this is easy oil, a day at the beach for you and your rough-neck crew.

But then as the oil is withdrawn, pressure drops, so back to work. Injecting water and gas, we are able to recover another 20–40% of the oil below. Finally, enhanced oil recovery (EOR), which injects heat or chemicals into the reservoir can eke out some of what remains. EOR oil is only around 2% of global supply (McGlade et al. 2018).

Oil accounts for a third of global primary energy and 95% of transportation energy. About 50% comes from just 100 giant oil fields—in the trade, they are known as "elephants"—and another 10% from 400 other giants, most of them discovered over 50 years ago, each with over half a billion barrels of oil. By the end of the oil age these fields will have produced 65% of all oil. Today the giants represent almost three-quarters of conventional oil reserves (Höök et al. 2009; Bai 2014; Doslson et al. 2019).

Conventional is usually "light" oil, with relatively short carbon chains and a high ratio of hydrogen to carbon. "Heavy" oil consists of Canada's tar sands, Venezuela's extra heavy oil, and other heavy oil found worldwide. This oil was ignored for a long time since it was often distant, expensive, low quality, hard to exploit and had little net energy return. This oil does not flow to the surface. You will not go to the beach. These fields require more time, capitalization, energy, expensive hydrogen-rich diluents, water, refining, and removal of toxic metals (Nikiforuk 2010; Faergestad 2016).

A. J. Friedemann, *Life after Fossil Fuels*, Lecture Notes in Energy 81, https://doi.org/10.1007/978-3-030-70335-6_2

Fig. 2.1 Spindletop 1901 and author Alice Friedemann enjoying a day at the beach

Now that cheap, easy conventional oil is harder to find, these tarry, heavy oils are being mined. Canada's tar sands comprise 10% of global oil reserves, dug out with 400 ton trucks and blasted with steam to get the 1–18% of bitumen oil from the tarry sludge. Another 17% of oil reserves are Venezuelan extra heavy oil. And finally 15% of remaining oil is heavy oil—worse than conventional, but less trouble than tar sands and extra heavy oil. So there is lots of oil left in the world, but it is not very good stuff.

There is a well-known saying in the oil industry: It is not the size of the tank, it is the size of the tap. The flow rate of heavy oil is a trickle compared to the gusher of conventional oil. Just 1.2 billion barrels, 3.4% of the 34.7 billion barrels of global oil production in 2019 came from tar sands and Venezuelan extra heavy oil (BP 2020; GOC 2020; Parashar 2019).

As stated earlier, the end of cheap and easy oil ended in 2005 when conventional oil plateaued, with production leveling off and declining since 2019. From now on the cost and difficulty of obtaining oil will increase. When the oil age began, the energy return on invested (EROI) was as high as 100:1, which means the energy of one barrel of oil could get 100 barrels more out of the ground. Today, conventional oil global EROI has shrunk to about 17:1, and in the US to about 11:1. Some scientists estimate an EROI of 10:1 or more is needed to keep modern society functioning (Hall and Cleveland 1981; Mearns 2008; Lambert et al. 2014; Murphy 2014; Fizaine and Court 2016; Hall 2017).

Canadian tar sand surface mining, which can extract 20% of tar sand reserves, has an EROI of 3.9–8. The other 80% can be extracted with in situ mining (drilling) at a lower EROI of 3.2 to 5.4:1 (Poisson and Hall 2013; Wang et al. 2017). Although Canadian tar sand production could be ramped up from the 1 billion barrels produced today to about to 2 billion barrels a year, production would still peak in 2040 (Söderbergh et al. 2007; NEB 2013).

What is saving the US now is unconventional, tight "fracked" oil. Fracking accounted for 63% of total US crude oil production in 2019 and 83% of global oil growth from 2009 to 2019 (Fig. 2.2). Fracked oil bought us time. However, there are compelling reasons why this unconventional oil remained in the ground until recent years. Due to the nature of fracking—fluids and sand are pumped under pressure to fracture shale and release trapped oil—costs are higher than conventional oil extraction. And unlike conventional oil fields, fracked fields have a much shorter lifespan. Once a fracked well begins to decline, it does so at a rate of about 82% over 3 years. Most investors in tight oil have not made any profits (Rowell 2020).

With most of the sweet spots drilled, high cost, bankruptcies, and lack of investment due to Covid-19, tight oil production may start declining in the mid-2020s (IEA 2018, Hughes 2019). Since 2015, 200 oil companies with over $130 billion in debt have gone bankrupt, and 100 more energy companies may fail in 2020 due to the pandemic and low prices (Wethe and Crowley 2020; KE 2020).

On a personal note, I would like to express my gratitude to all those hard luck fracked oil investors. They may have thrown good money after bad, had to do without his and her yachts, but not for naught. The fracked oil bought us time.

Other nations are not getting on the fracking bandwagon, and there are good reasons why. Most tight shale oil is produced in the US because of existing,

World oil production fell -9.7 mmb/d in May 2020 & another -1.1 mmb/d in June
Output had fallen -2.1 mmb/d from November 2018 peak *before* the May collapse
U.S. tight oil accounted for 83% of global oil growth from 2009 through 2019

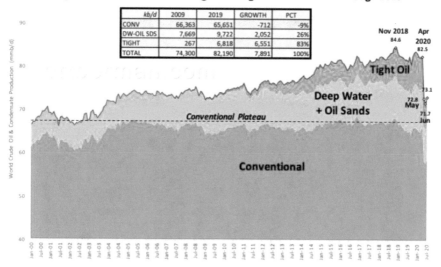

kb/d	2009	2019	GROWTH	PCT
CONV	66,363	65,651	-712	-9%
DW-OIL SDS	7,669	9,722	2,052	26%
TIGHT	267	6,818	6,551	83%
TOTAL	74,300	82,190	7,891	100%

Fig. 2.2 Output expected to be flat through 2021 from 10.5 to 11 mmb/d. US tight oil accounted for 83% of growth in world production from 2009 to 2019. Deep water and oil sands were the other growth area at 23% while conventional production declined 9% over the same period. Source: Berman (2020)

widespread oil infrastructure, half the world's drilling rigs, subsurface mining rights, nearby customers, a workforce with years of experience, and easy financing.

Conventional crude oil production leveled off in 2005, and it appears to have peaked in 2008 at 69.5 million barrels per day (mb/d) according to Europe's International Energy Agency (IEA 2018 p45). The US Energy Information Agency shows global peak crude oil production at a later date in 2018 at 82.9 mb/day (EIA 2020) because they included tight oil, oil sands, and deep-sea oil.

This does not mean we have reached peak oil production for certain. The 2018 decline coincided with OPEC production cuts. We will not know for years, in hindsight. The exact year does not matter. What does matter is that it will take decades to scale up renewables, which require oil to be constructed and maintained, and we are running out of time.

The IEA forecast a supply crunch by 2025 in their rosy New Policies scenario, which assumes greater efficiencies and alternative fuels are adopted (Fig. 2.3). By 2025, with 81% of global oil declining at up to 8% a year (Fustier et al. 2016; IEA 2018), 34 mb/day of new output will be needed, and 54 mb/day if facilities are not maintained. That is more than three times Saudi Arabian production. The 15 mb/day of predicted US shale is not likely since the IEA shows it declining in the mid-2020s (IEA 2018 Table 3.1).

The IEA (2018) points out that the new conventional crude projects approved over the last 3 years are only half the amount needed by 2025. With Covid-19, few projects are likely to begin. Oil prices have gone down so much that exploration and production investments were plummeting before Covid-19 and many fields already found are too costly to develop. Already 4 years of world oil consumption, 125 billion barrels, are likely to be written off by oil companies and remain in the ground (Rystad 2020b; Hurst 2020).

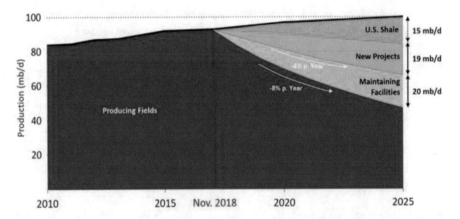

Fig. 2.3 Expected decline of oil production (red area) and possible stopgaps to the decline rate of 4–8% a year. Maintaining facilities refers to enhanced recovery from existing fields. (IEA 2018). (Modified figure 1.19 from IEA)

Globally, new oil discoveries have fallen for 6 years, with consumption of oil six times greater than discoveries from 2013 through 2019 (Rystad 2020a; BP 2020). And in the US, conventional oil and gas discoveries from 2016 to mid-2020 were at their lowest levels for 70 years (BW 2019).

This should not be a surprise. We have been bellied up to the bar, drinking barrels of the crude elixir, for a long time. Since 1950, we have consumed about 1.5 trillion barrels, about half of the easy, conventional oil. Now the remaining oil is increasingly of lower quality and expensive because it is harder to come by, or requires EOR. These deposits are locked in oil sands, tight shale deposits, extra heavy oil deposits, deep under the ocean, or in the Arctic where permafrost on land and icebergs make it difficult and costly to obtain.

But It Ain't over Until the Fat Lady Sings

Perhaps more oil can be produced with new enhanced or improved oil recovery methods, though that can cause a steeper decline rate after the peak is reached (Kuishuang et al. 2013). More oil could also be extracted if offshore oil platforms were invented that could dodge hurricanes and icebergs.

Non-geological factors can keep oil flowing too: A booming global economy. World peace, especially in the Middle East where half of the global oil reserves are. Oil-producing nations continuing to export oil, despite increasing consumption from their own growing populations.

Although we may only have used about half the world's oil (BP 2020), much will remain underground, unrecoverable. As it is, only 35% of an oil field is produced on average, the rest remaining underground (Maugeri 2009). With the remaining reserves increasingly heavy, in deep water offshore, or arctic oil, what remains may take centuries to obtain. Since it has such a low energy return, more energy will be used to get it, which means less energy for society's other needs. Or, because of its EROI of less than 10:1, this unconventional oil may forever remain off-limits.

The time we once had to do research on fossil fuel alternatives and scale them up to commercial level is running out. With the demise of the Age of Easy Cheap Oil, our day at the beach is ending.

References

Bai G (2014) Giant fields retain dominance in reserves growth. Oil Gas J 112:44–51
Berman A (2020) Stop expecting oil & the economy to recover. https://www.artberman.com/2020/09/03/stop-expecting-oil-and-the-economy-to-recover/#:~:text=Global%20oil%20markets%20are%20on,through%20to%20the%20next%20century. Accessed 1 Nov 2020
BP (2020) Statistical review of world energy. British petroleum
BW (2019) Conventional discoveries have fallen to lowest levels in 70 years. Business Wire. https://www.businesswire.com/news/home/20191001006010/en/. Accessed 1 Nov 2020

Doslson JC, Merrill R, Sternbach C et al (2019). Advances in stratigraphic trap exploration. GEOExPro

EIA (2020) International Energy Statistics. Petroleum and other liquids. Data Options. U.S. Energy Information Administration. Select crude oil including lease condensate to see data past 2017

Faergestad IM (2016) The defining series: heavy oil. Oilfield Review, Schlumberger. https://www.slb.com/resource-library/oilfield-review/defining-series/defining-heavy-oil. Accessed 1 Nov 2020

Fizaine F, Court V (2016) Energy expenditure, economic growth, and the minimum EROI of society. Energy Policy 95:172–186

Fustier K, Gray G, Gundersen C, et al (2016) Global oil supply. Will mature field declines drive the next supply crunch? HSBC Global Research

GOC (2020) Crude oil facts. Oil sands. Government of Canada. https://www.nrcan.gc.ca/science-data/data-analysis/energy-data-analysis/energy-facts/crude-oil-facts/20064. Accessed 1 Nov 2020

Hall CAS (2017) Energy return on investment: a unifying principle for biology, economics and sustainability. Springer Nature, New York

Hall CAS, Cleveland CJ (1981) Petroleum drilling and production in the United States: yield per effort and net energy analysis. Science 211:576–579

Höök M, Hirsch R, Aleklett K (2009) Giant oil field decline rates and their influence on world oil production. Energy Policy 37:2262–2272

Hughes D (2019) Shale reality check 2019. Post Carbon Institute

Hurst L (2020) Oil companies wonder if looking for oil is still worth it. https://www.rigzone.com/news/wire/oil_companies_wonder_if_looking_for_oil_is_still_worth_it-17-aug-2020-163038-article/. Accessed 1 Nov 2020

IEA (2018) International Energy Agency World Energy Outlook 2018, figures 1.19 and 3.13. International Energy Agency

KE (2020) 19 energy companies have filed for bankruptcy in 2020: law firm. https://www.kallanishenergy.com/2020/06/05/19-energy-companies-have-filed-for-bankruptcy-in-2020-law-firm/. Accessed 1 Nov 2020

Kuishuang F, Kerschner C, Hubacek K (2013) Economic vulnerability to peak oil. Glob Environ Chang

Lambert JG, Hall CAS, Balogh S et al (2014) Energy, EROI and quality of life. Energy Policy 64:153–167

Maugeri L (2009) Squeezing more oil out of the ground. Scientific American

McGlade C, Sondak G, Han M (2018) Whatever happened to enhanced oil recovery? International Energy Agency

Mearns E (2008) The global energy crisis and its role in the pending collapse of the global economy. Presentation to the Royal Society of Chemists, Aberdeen, Scotland. http://www.theoildrum.com/node/4712

Murphy DJ (2014) The implications of the declining energy return on investment of oil production. Phil Trans R Soc A. https://doi.org/10.1098/rsta.2013.0126

NEB (2013) Canada's energy future, energy supply and demand to 2035. National Energy Board, Government of Canada

Nikiforuk A (2010) Tar sands: dirty oil and the future of a continent. Greystone books

Parashar M (2019) Venezuela's Orinoco Belt crude production falls to 246,000 b/d: technical report. S&P Global Platts. https://www.spglobal.com/platts/en/market-insights/latest-news/oil/092419-venezuelas-orinoco-belt-crude-production-falls-to-246000-b-d-technical-report. Accessed 1 Nov 2020

Poisson A, Hall CAS (2013) Time Series EROI for Canadian Oil and Gas. Energies 6:5940–5959

Rowell A (2020) With U.S. shale peaking & having "never made money," investors lose billions. http://priceofoil.org/2020/06/26/with-u-s-shale-peaking-having-never-made-money-investors-lose-billions/. Accessed 21 Nov 2020

Rystad (2020a) Global oil and gas discoveries reach four-year high in 2019, boosted by ExxonMobil's Guyana success. https://www.rystadenergy.com/newsevents/news/press-releases/global-oil-and-gas-discoveries-reach-four-year-high-in-2019/. Accessed 1 Nov 2020

Rystad (2020b) Covid-19 report. Global outbreak overview and its impact on the energy sector. Rystad Energy

Söderbergh B, Robelius F, Aleklett K (2007) A crash programme scenario for the Canadian oil sands industry. Energy Policy 35

Wang K, Vredenburg H, Wang J, et al (2017) Energy return on investment of Canadian Oil Sands extraction from 2009 to 2015. Energies. https://doi.org/10.3390/en10050614

Wethe D, Crowley K (2020) Shale's bust shows basis of boom: Debt, debt, and debt. https://www.washingtonpost.com/business/energy/shales-bust-shows-basis-of-boom-debt-debt-and-debt/2020/07/22/0e6ed98c-cc41-11ea-99b0-8426e26d203b_story.html. Accessed 1 Nov 2020

Chapter 3
When the World Ran on Wood

Before fossil fuels, civilizations were powered by wood energy and built with wood infrastructure. Biomass fed and fueled the muscle power of animals and people.

Going back to our beginnings, burning biomass may be why we evolved from apes into homo sapiens. Fire provided warmth, made food safe, made indigestible or poisonous food edible, and reduced spoilage. Fire kept predatory animals at bay and allowed us to migrate to colder climates. Cooked food has more calories than raw food, helping our brains to evolve to triple in size and subsist on a few meals instead of eating leaves all day long (Wrangham 2009). Best of all, the warmth and light gave us more time to socialize and develop language. It would seem the barbecue predates the Weber grill!

In Wood World, charcoal from wood brewed beer, smelt metals, made steel, weapons, jewelry, ceramics, glass, and bricks. Wood was the indispensable substance of ships, homes, fortresses, log corduroy roads, bridges, wheels, bowls, wagons, furniture, tools, crates, barrels, fences, utensils, docks, musical instruments, paper, and more.

Forests and the Rise and Fall of Civilizations

Since wood was the sine qua non of civilization, nations supplemented domestic lumber supplies by building enormous wooden merchant and warships. These ships sailed to faraway lands to trade, or wage war, for wood, as well as for grain, wine, spices, and textiles.

Nations collapsed after being logged out and finding themselves unable to conquer or trade with forested nations (Cline 2014; Diamond 2005; Perlin 2005). Deforestation also increased topsoil erosion, another major reason civilizations collapsed (Montgomery 2007; Scharf 2012).

© The Author(s), under exclusive license to Springer Nature Switzerland AG 2021
A. J. Friedemann, *Life after Fossil Fuels*, Lecture Notes in Energy 81,
https://doi.org/10.1007/978-3-030-70335-6_3

The Roman empire far outstripped its ability to feed the million people inhabiting Rome, and became dependent on food shipped from Carthage. But when barbarians captured Carthage in the fourth century it was a major reason for the fall of the Roman Empire (Heather 2009).

Transportation in Wood World

Before fossil fuels, shipping was faster and cheaper than land transport, so nearly everyone lived near navigable water (Fang and Jawitz 2019).

Adam Smith, the famous economist, estimated 200 tons of goods carried the 375 miles over land from London to Edinburgh required the labor of 100 men for 3 weeks, as well as 400 horses and 50 large wagons. The same quantity of goods carried by ship required just six to eight men, with little wear and tear on the ship, whereas the wagons could be lost fording rivers and damaged from rough roads (Wrigley 2010).

Oxen and horses were limited in what they could haul by the need for their own feed to be carried. It took a ton of hay to feed two oxen for a week of travel, a fifth of the cargo.

There were few roads on which to haul cargo. What roads existed were near towns, full of potholes, and deeply rutted, forcing most overland goods to be carried on pack animals or by humans on rough paths for most of the journey.

Rome changed that by building the first advanced road system, with 250,000 miles of roads, 50,000 of them paved, a network unmatched until the nineteenth century. These roads were mainly built to quickly deploy Roman armies and haul food to cities and armies on the move.

Land and Forests Restricted the Size of Cities

Before fossil fuels, societies were able to make their forests last longer than today. Felling tall trees and killing them was rare except for special needs such as making bridges or ships. For firewood and other needs, trees were cut in a way that encouraged new shoots to sprout that could be harvested every few years. Coppiced forests were more biodiverse than today's plantations since many kinds of wood were planted, each kind suited to different purposes.

Coppiced firewood needed to be harvested within 9–18 miles (15–30 km) of where they were used since the wood was hauled on carts over bad roads. Beyond 18 miles the energy content of the wood was less than the energy of the pasture used to feed the horse to make the round trip (Sieferle 2001).

Even so, it did not take long for forests to disappear. A typical home burned from 1 to 1.6 tons of firewood a year requiring roughly 1.6 acres of forest a year, and a city with half a million people needed 800,000 acres of woods weighing 650,000 tons a year. Hauling this wood to the city would have required 500,000–750,000 horse-drawn carts (Wrigley 2010). Perlin gives examples of nearly complete deforestation of Peloponnesia and many other regions over the millennia.

For most of human history, the size of cities was limited by how heavy and expensive it was to transport wood from nearby forests. Cities had to draw on nearby areas at least 30 times their size for fuel supply (Smil 2017). So, today's Chicago, with 2.7 million people spread over 234 square miles, would need to expand to 7000 square miles, raze all the suburbs where 6.8 million people now live, and plant trees.

Charcoal would have required even more land than that. Large preindustrial cities in a northern temperate climate that relied heavily on charcoal would have needed a wooded area at least 100 times their size to ensure a continuous supply. This reality restricted the growth of cities even where food and water were abundant (Smil 2017).

Consequently, for most of our history, there was very little population growth. Economists knew that to create wealth, there was only labor, capital, and land. It was possible at times to expand labor and capital. But the supply of land was fixed. Any attempt to grow society meant more people needed to be fed. But that could not be done if available land was poor or degraded for crops or dominated by non-agricultural uses. Growing crops on inferior land required more labor and capital with a lower rate of return. At some point back before fossil fuels, limited land put a cap on growth, regardless of how clever people were, and regardless of their political, social, and economic systems.

Wood provided thermal power, while muscle power was the source of mechanical energy, but limited to a fraction of the calories consumed. That is because we warm-blooded creatures use such a large part of food intake for basic body maintenance. It takes about 1500 kilocalories a day to stay alive even if no work is performed. With 2500 kilocalories, only 40% of calories are available for productive work, and it takes 3500 to double the work done. The same is true of oxen and horses. Ill-fed animals will use a high proportion of their food intake to stay alive, leaving only a small proportion of energy to drag a plow or pull a cart.

A well-fed horse can do six times as much work as a man. But just to replace current US farm machinery with draft animals would require over 200 million horses fed on 1.2 million square miles of land, twice the area of US arable land (Smil 2017). The good news: There would be enough manure for every garden and farm in the world!

Fossil World

London's population exploded eightfold from 60,000 in 1534 to 530,000 in 1696. Consequently, deforestation from agriculture, heat, and cooking was so extensive in the sixteenth century that the British were forced to use coal, launching the industrial revolution and our lifestyles of today (Nef 1977).

Despite lots of coal for thermal energy, the wood for infrastructure and the ships that maintained the British Empire were scarce. So much land was being cleared for agriculture that reforestation was not happening, so Britain imported wood, mainly from the American colonies and the Baltic region.

By 1800, global coal production had reached over 57 million barrels of oil equivalent (BOE). Each barrel contains the work of 4.5 years of manual labor a man can perform (Hagens and White 2017). For the previous 400 years, about a third to half of all work went to obtaining food, firewood, and animal feed, implying an EROI of 2 or 3 to 1. But when fossil fuels arrived, the EROI jumped to 20:1 or more and less than 10% of activities had to be devoted to getting the energy required to run society (Hall 2017).

In 1920, natural gas and oil extraction began in earnest worldwide. Today 31.2 billion barrels of oil, 25.4 BOE coal, and 21 billion BOE of natural gas are extracted, for a grand total of 77.6 billion BOE every year (Smil 2016; BP 2017), doing work it would take 3.4 trillion men and women to do in a year.

Fossil fuels can do what humans or wood cannot, providing temperatures of up to 3000 Fahrenheit to smelt metals and make glass, bricks, ceramics, and more. The density of energy embodied in oil is difficult to appreciate. If you have ever had to push a broken-down car to the side of the road, think about your mighty exertions and then consider this humbling comparison: Fossil fuels propel four-ton cars, 448-ton Boeing 747–8 freight airplanes, 13,000-ton freight trains, and 100,000-ton container ships … for thousands of miles. Each locomotive engineer controls energy equivalent to that of 100,000 men; each jet pilot the energy equivalent of 700,000 men (Rickover 1957). In our world today, we cross the oceans in hours rather than weeks or months. We fly 30,000 ft above the planet below. We live more like Gods than Kings.

Welcome to Fossil World, where lifespans are twice as long as in Wood World. No longer do up to 90% of us have to farm for the fortunate 10% who live in towns and cities (Smil 2017). Lights blaze, temperatures are always comfortable, we vacation around the world, and have endless entertainment on computers and TV. There are hundreds of millions of products to shop for that would fill a catalog many miles wide (Amazon has 350 million items for sale—all ready to be delivered to your doorstep). We take this for granted but an ancient king would be amazed at our personal affluence.

Fossils allowed us to break out of the prison of land, labor, and capital limits to growth and expand our economy and population exponentially for 200 years. But there still are limits to growth. Alas, fossil fuels are finite.

References

BP (2017) BP statistical review of world energy. British Petroleum
Cline EH (2014) 1177 B.C. The year civilization collapsed. Princeton University press
Diamond J (2005) Collapse. How societies choose to fail or succeed. Viking Press

Fang Y, Jawitz JW (2019) The evolution of human population distance to water in the USA from 1790 to 2010. Nat Commun 10:430

Hagens N, White DJ (2017) GDP, jobs, and fossil largesse. https://www.resilience.org/stories/2017-11-30/gdp-jobs-and-fossil-largesse/. Accessed 1 Nov 2020

Hall CAS (2017) Will EROI be the primary determinant of our economic future? The view of the natural scientist versus the economist. Joule 1:635–638

Heather P (2009) Empires and barbarians: the fall of Rome and the birth of Europe. Oxford University Press

Montgomery DR (2007) Dirt: the erosion of civilizations. University of California Press

Nef JU (1977) An early energy crisis and its consequences. Sci Am

Perlin J (2005) A Forest journey: the story of wood and civilization. Countryman Press

Rickover HG Admiral (1957) U.S. Navy. Energy Resources and Our Future. Scientific Assembly of the Minnesota State Medical Association. http://large.stanford.edu/courses/2011/ph240/klein1/docs/rickover.pdf. Accessed 14 Nov 2020

Scharf P (2012) Erosion and the value of topsoil. The long view. University of Minnesota

Smil V (2016) Energy transitions: global and national perspectives. Praeger

Smil V (2017) Energy and civilization: a history. The MIT press

Sieferle RP (2001) The Subterranean Forest: Energy Systems and the Industrial Revolution. White Horse Press

Wrangham R (2009) Catching fire: how cooking made us human. Basic Books

Wrigley EA (2010) Energy and the English industrial revolution. Cambridge University Press

Chapter 4
We Are Alive Thanks to Fossil-Fueled Fertilizer

Only the uncontacted people in remote areas like the Amazon rainforest are untouched by fossil fuels.

For those in the US, natural gas provides much of our comfort for heating, cooking, and refrigeration. This is how a third of US natural gas is used. Another third generates electricity for lighting, TV, air-conditioning, computer, and other devices, with most of the rest for manufacturing cement, steel, millions of products, and fertilizer (EIA 2019).

Natural Gas Fertilizer and the Population Explosion from 1.6 to 7.8 Billion People

Next time you walk into a supermarket, put your craving for fresh strawberries on hold, stanch your yearning for ice cream, stifle your desire for steak. Instead, think about fertilizer. Let us not be ingrates: At least four billion of us are alive due to fertilizer (ammonia) made by the high heat and pressure generated by natural gas heat to force nitrogen in the air to combine with the hydrogen in natural gas (Fisher 2001; Smil 2004; Stewart et al. 2005; Erisman et al. 2008). This allows up to five times more food to be grown than would otherwise be the case. For example, 8700 pounds of corn can be harvested per acre today; just 1600 pounds in the past before natural gas fertilization (Schulz 2007).

In the past, nature replenished nitrogen slowly after a field was left fallow for a year. Then many improvements were made so that land was not left fallow. Manure, compost, and legumes added nutrition and fixed nitrogen in the soil. Crops were rotated. Production per acre increased dramatically. But nothing like with artificial ammonia.

Guano, the accumulated excrement of seabirds and bats that is loaded with nitrogen, phosphate, and potassium, once was a prized commodity. Islands with guano

© The Author(s), under exclusive license to Springer Nature Switzerland AG 2021
A. J. Friedemann, *Life after Fossil Fuels*, Lecture Notes in Energy 81, https://doi.org/10.1007/978-3-030-70335-6_4

were discovered, and in the nineteenth century, wars were fought over guano islands in South America. Spanish troops actually occupied the guano-rich Chincha Islands of Peru from 1864 to 1866. Hey soldier, guard that guano with your life. That is service above and beyond the call of duty!

It was not long before the guano was depleted, yet the global population was still growing. The world needed a new source of nitrogen. Famine was averted when two German scientists, Bosch and Haber, invented a way to make ammonia from the nitrogen in the air and the hydrogen in fossil fuels, at first using coal and later natural gas. By the end of WWI fertilizer added more nitrogen to the soil than had ever been possible using natural sources. Hail technology! Guano was no longer worth fighting over.

Thanks to coal, the population had risen from 500 million in the 1600s to 1.6 billion in 1910 and hit the wall until the Haber-Bosch fertilizer process came along, greatly increasing the human carrying capacity of the earth for four billion more people. Add a billion more due to pesticides, herbicides, insecticides, and fungicides made from natural gas, oil, or coal. And another billion owe their lives to the "Green Revolution" of plants bred for higher yields, phosphorus fertilizer, as well as fossil-fueled irrigation, farm management technology, diesel-powered agricultural equipment, food distribution, and processing. So here we are, with over seven billion of us alive thanks to fossil fuels, and three billion more expected onboard by 2050.

You may not drink petroleum, and yet you do. Fossil fuels are required for every step of your meal, from planting, fertilizing, harvesting, distributing, processing, packaging, refrigerating, to cooking at home. About 10 kcal of fossils are required to produce one kilocalorie of food by the time it hits your plate. All this sort of takes your appetite away, yes? (Fig. 4.1).

Fertilizer energy is 28% of the energy used in agriculture (Heller and Keoleian 2000), because making ammonia fertilizer from natural gas, with natural gas as the energy source, requires a lot of heat, up to 932 °F, and pressure up to 3600 pounds per square inch. That is a lot of pressure. If you were turned into fertilizer, your systolic blood pressure of 120 would be on the high side, about 186,000.

These high pressures and temperatures are needed to force introverted triple bonded nitrogen (N_2) out of the air—it would rather not interact with anything—and to drive it to combine with hydrogen from natural gas (CH_4) to make ammonia (NH_3).

Nitrogen is vital because it is often the factor limiting crop growth since it is used in photosynthesis and in proteins, DNA, RNA, muscle, and hence milk, eggs, and meat.

The use of fertilizer has grown exponentially. More than half of all synthetic nitrogen fertilizers ever used on the planet have been spread across fields since 1985 (UN 2005). This can not go on forever, since conventional natural gas production peaked in the United States in 1973 and has been declining at a rate of 5% a year since then. What is saving us now is tight and shale gas, but that too has a dismal future, due to low prices, Covid-19, and bankruptcies. Once a fracked natural gas well begins to decline it does so at a rate of about 82% over 3 years. The IEA (2018, 2020) estimates natural gas may start to decline by 2035, or sooner since 30% comes from fracked oil wells that may start to decline in the mid-2020s.

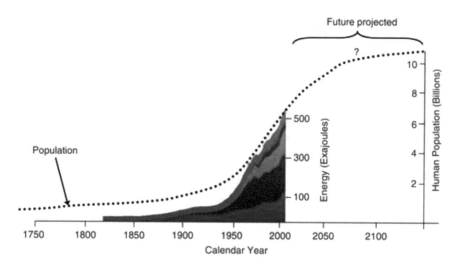

Fig. 4.1 Energy use and human population growth from 1750. Utilization of these energy sources, together with the energy used by humans from net primary production, is now approaching the entire energy available to the global ecosystem before human intervention (Barnosky 2015). Blue = coal; red = oil; green = natural gas; purple = nuclear; light blue = hydro; orange = biomass. (Source: Williams et al. (2016) and http://www.theoildrum.com/node/8936)

When US fracked gas runs low, we could keep the party going by importing natural gas on ships in the form of liquefied natural gas (LNG), chilled to −256 F on enormous ships. Tens of billions of dollars would be required to build LNG import infrastructure. No wonder 90% of the world's natural gas travels in much cheaper pipelines.

In the US, we have done the opposite and built export facilities to send our finite natural gas to other countries (Chapa 2018). We are now the fourth-largest LNG exporter in the world (Statista 2020). Doh!

Fertilizer Harms the Land and Atmosphere

Fertilizer has its downside. Fertilizer releases nitrous oxide (N_2O), a greenhouse gas with a global warming potential 300 times greater than carbon dioxide. N_2O is also the largest destroyer of stratospheric ozone (Ravishankara et al. 2009), which shields plants and animals from the damaging ultraviolet light (UVB) that reduces crop productivity, photosynthesis, and quality while increasing their susceptibility to disease. Worst of all, UVB harms phytoplankton, which produces half of all oxygen and are the main absorbers of greenhouse gas carbon, which they sequester on the ocean floor when they die. They are also at the bottom of the ocean food web. Nitrogen runoff is especially harmful to freshwaters, accelerating eutrophication.

The agricultural sector accounts for 73% of N_2O emissions (EIA 2011). When too much fertilizer is applied, the microbes go to town and emit exponentially more N_2O (Shcherbak et al. 2014).

Fertilizer also harms the soil ecosystem. A balanced diet for soil organisms is about 20 parts carbon to one part nitrogen. Too much nitrogen and too little carbon starves and eventually kills them. The beneficial functions microbes perform for plants, such as defending crops from pests and diseases, also are lost, so farmers add even more fertilizer and pesticides.

I do not mean to be a nag, but can we get farmers to stop using so much fertilizer made from finite natural gas? Although some farmers are doing so, 40% of farms are leased. The more crops that renting farmers can grow, the more money they and the owners make, which leads to over-fertilization. This problem is exacerbated by government commodity payments. Meant to be a safety net, these subsidies are often used instead to buy and apply excess fertilizer that pollutes rivers, lakes, and oceans, increasing water treatment and health costs and killing fish (Broussard et al. 2012, Ma et al. 2012; Troeh and Thompson 2005).

What Could Replace Fertilizer? The Dirt on Dirt

Since the dawn of agriculture, biomass from crop residues and animal manure was composted by farmers to make fertilizer. Now we call it organic farming. Compost not only fertilizes, it creates a soil ecosystem that protects plants from diseases and pests, and aerates the soil to allow water and air to reach the roots. Plants need to drink and breathe just like we do. Compost also protects the soil by storing water, saving the farmer money and time by reducing the need to pump irrigation or vanishing aquifer water. More carbon is stored as well.

On organic farms and wildlands, the land is alive. If the trillions of microorganisms in the earth had voices the hills would be alive with the sound of music. There is so much life in the soil, there might be 10 "biomass horses" below for every horse grazing above on an acre of pasture (Hemenway 2009). Just a teaspoon of the earth has millions of species and—see if you can get your head around this—more organisms than people on earth (Bogard 2017; Wardle 2004). If you dove into the soil and swam around, you would see plant roots that looked like upside-down trees and be surrounded by thousands of miles of thin strands of mycorrhizal fungi that help plant roots absorb nutrients and water (Pennisi 2004).

You could think of the soil biota as tiny engineers, sequestering carbon, filtering pollutants, lowering greenhouse gas emissions, structuring the soil to hold more water to see plants through droughts and dry spells, fixing nitrogen (fertilizer) naturally in the soil, and constructing underground tunnels for air and water to infiltrate. They provide plants an immune system against diseases, weeds, and insects and break down dead vegetation into nutrients plants can absorb to complete the circle of life.

Looking ahead to the day there is not enough fossil energy, organic agriculture which does without fossil fuel-based fertilizers, can mitigate our path forward. If we want to grow more biomass for transportation, manufacturing, and to feed 10 billion people in 2050, then returning to organic agriculture and improving the soils must be our top priority.

It is ironic that fossil fertilizers and pesticides destroy soil biota as well as accelerate nutrient loss from topsoil via erosion. Montgomery (2007) says that "this puts us in the odd position of consuming finite fossil fuels—geologically one of the rarest and most useful resources ever discovered—to provide a substitute for dirt, the cheapest and most widely available agricultural input imaginable."

References

Anthony D. Barnosky, (2015) Transforming the global energy system is required to avoid the sixth mass extinction. MRS Energy & Sustainability 2 (1).

Barnosky AD (2015) Transforming the Global Energy System Is Required to Avoid the Sixth Mass Extinction. MRS Energy Sustainability.

Bogard P (2017) The ground beneath us. From the oldest cities to the last wilderness, what dirt tells us about who we are. Little, Brown & Co.

Broussard WP, Turner RE, Westra JV (2012) Do federal farm policies influence surface water quality? Agric Ecosyst Environ 158:103–109

Chapa S (2018) LNG tanker arrives at Cheniere Energy's Corpus Christi plant. Houston Chronicle

EIA (2011) Emissions of greenhouse gases in the U.S. U.S. Energy Information Administration

EIA (2019) Natural gas explained. Use of natural gas. U.S. Energy Information administration

Erisman JW, Sutton MA, Galloway J, et al (2008) How a century of ammonia synthesis changed the world. Nat Geosci

Fisher D (2001) The Nitrogen Bomb. By learning to draw fertilizer from a clear blue sky, chemists have fed the multitudes. Discover magazine

Heller MC, Keoleian GA (2000) Life-Cycle Based Sustainability Indicators for Assessment of the U.S. Food System. University of Michigan: Ann Arbor

Hemenway T (2009) Gaia's Garden: A Guide to Home-Scale Permaculture, 2nd Edition. Chelsea Green Publishing

IEA (2018) The future of petrochemicals towards more sustainable plastics and fertilisers. International Energy Agency

IEA (2020) Gas 2020. Analysing the impact of Covid-19 on global natural gas markets. International Energy Agency

Ma S, Swinton SM, Lupi F et al (2012) Farmers' willingness to participate in payment-for-environmental-services programmes. J Agric Econ 63:604–626

Montgomery DR (2007) Dirt: the erosion of civilizations. University of California Press

Pennisi E (2004) The secret life of fungi. Science 304:1620–1623

Ravishankara AR, Daniel JS, Portmann RW (2009) Nitrous oxide (N2O): the dominant ozone-depleting substance emitted in the 21st century. Science 326:123–125

Schulz WG (2007) The costs of biofuels. Chemical and Engineering News

Shcherbak L, Millar N, Robertson GP (2014) Global metaanalysis of the nonlinear response of soil nitrous oxide (N2O) emissions to fertilizer nitrogen. Proceedings of the National Academy of Sciences. https://doi.org/10.1073/pnas.1322434111

Smil V (2004) Enriching the earth: Fritz Haber, Carl Bosch, and the transformation of world food production. MIT Press

Statista (2020) World's leading gas exporting countries in 2019. Statista.com

Stewart WM, Dibb DW, Johnston AE et al (2005) The contribution of commercial fertilizer nutrients to food production. Agron J 97:1–6

Troeh F, Thompson LM (2005) Soils and Soil Fertility, 6th edition. Wiley-Blackwell

UN (2005) Millennium ecosystem assessment synthesis report. United Nations

Wardle D (2004) Ecological linkages between aboveground and belowground biota. Science 304:1629–1633

Williams M, Zalasiewicz J, Waters CN, et al (2016) The Anthropocene: a conspicuous stratigraphical signal of anthropogenic changes in production and consumption across the biosphere. Earth's Future

Chapter 5
Without Transportation, Civilization Ends

Wood World Horses Are Now Fossil World Diesel Engines and Gas Turbine Machines

Let us hear it for man's best workhorse, the horse. In Wood World, they did the hard work of civilization for thousands of years—plowing, hauling goods, moving armies, construction work, transportation, and more. If one horsepower equals the work of one horse in a day and an average car has 180 horsepower and the world has 1.4 billion cars, then we have got at least 252 billion horses running around, most doing little real work (Autolist 2019; Chesterton 2018). Add up the horsepower of trucks, ships, locomotives, airplanes, and equipment and there are trillions of invisible horses out there.

Diesel trucks began to replace horses in the 1930s and now we have all kinds of trucks: tractors, harvesters, 18-wheelers, garbage, cement, mining, fire, tanks, logging, bulldozers, landfill compactors, pipelayers, excavators, asphalt pavers, scrapers, trenchers, and cranes that can reach to the sky and lift building supplies and segments of 2000-ton wind turbines skyscraper high.

Ships carry 90% of all traded goods. Locomotives can pull over a 100 railcars. Yet trucks are the most important. Look around you. Find one thing that was never on a truck … I am waiting.

The fact is, virtually all goods were on a truck at some point, if only for delivery or to a port or rail yard. In the US, trucks deliver 80% of goods over 4.1 million miles of roads, with 80% of towns completely dependent on trucks. Ships roam a mere 25,000 miles of navigable ocean, river, and lake waterways while freight trains have just 95,000 miles of tracks.

What do trucks, trains, and ships have in common? They all are powered by petroleum, almost always diesel. Including 97% of class 8 through class 13 trucks (MDOT 2012).

© The Author(s), under exclusive license to Springer Nature Switzerland AG 2021
A. J. Friedemann, *Life after Fossil Fuels*, Lecture Notes in Energy 81,
https://doi.org/10.1007/978-3-030-70335-6_5

When Trucks Stop Running

If diesel fuel suddenly became unavailable, civilization would end within a week. When trucks stop running, here is how it would play out:

Day 1: Trucks stop running.

Day 2: Most groceries vanish from shelves, and the multiple daily deliveries of produce, meat, beer, chilled goods, bread, and dairy products stop.

Day 3: Almost no businesses carry inventory because they depend on just-in-time delivery, so hospitals, pharmacies, restaurants, and construction slow down and start to run out of supplies.

Day 4: Garbage begins to pile high, sewage treatment sludge and slime tanks fill up, mass transit stops, layoffs are ubiquitous. Ambulances, police, and fire services last as long as private fleets have diesel.

Days 5–60: Workers are laid off, ATMs stop dispensing cash, service stations are out of fuel, 685,000 tons of trash are added every day in the US. Products stop being manufactured. If it is harvest time, crops rot in the field. Coal power plants run out of coal, water treatment plants out of purification chemicals (Holcomb 2006; McKinnon 2004; SARHC 2009).

Day 61: You are told not to worry. Everything that you are observing is not as it seems!

Since Diesel Is Finite, We Need to Replace It with Something Else

Diesel performs unappreciated, unrecognized daily miracles. It is the ultimate essential worker, and it will take a miracle worker to replace diesel. Diesel has the highest energy density of any fuel other than uranium and is essential to move 80,000-pound trucks while lifting, digging, or traveling hundreds of miles. Just 1 L (four cups) of diesel can move one of these very heavy trucks for 2 miles on level ground.

Diesel engines have been refined for over a century to burn diesel efficiently, and can not operate on gasoline, ethanol, natural gas, hydrogen, or diesohol. They last up to 40 years, are twice as energy efficient as gasoline engines, and deliver far more horsepower and torque to do the hard work that keeps civilization running.

Diesel will not suddenly disappear, and peak oil does not mean "running out of oil." It just means that global oil production has reached peak production, and will then decline. So, what could replace diesel?

References

Autolist (2019) What is the Average Car Horsepower? https://www.autolist.com/guides/average-car-horsepower. Accessed 1 Nov 2020

Chesterton A (2018) How many cars are there in the world? Short answer? Lots. Lots and lots and lots. https://www.carsguide.com.au/car-advice/how-many-cars-are-there-in-the-world-70629. Accessed 1 Nov 2020

Holcomb RD (2006) When trucks stop, America stops. American Trucking Association

McKinnon A (2004) Life without Lorries: The impact of a temporary disruption of road freight transport in the UK. Commercial Motor Magazine

MDOT (2012) Definitions of visual classification of heavy commercial vehicles by body type, industry, and any additional details. Minnesota Department of Transportation

SARHC (2009) A week without truck transport. Four Regions in Sweden 2009. Swedish Association of Road Haulage Companies

Chapter 6
What Fuels Could Replace Diesel?

Seeing a future where oil reserves would inevitably decline, the Department of Energy (DOE 2002) began researching the question of what would replace oil many years ago. They concluded that there was no energy source that could substitute for liquid hydrocarbon fuels because nothing else is so abundant, has such high energy and power density, stores energy so efficiently and conveniently, releases stored energy so readily, has an existing infrastructure, is so easily transported, and has the requisite volume. They concluded that hydrocarbon-based fuels would be the future fuel of heavy-duty vehicles for the next 25 years.

Here we are 18 years later. No energy source has emerged to rival or replace hydrocarbon fuels.

Wanted: A Renewable Commercial Fuel for Existing Engines and Infrastructure

Clearly, the optimal fuel would be one that billions of combustion engines could use without modifications or entirely new engines. This is the holy grail, a "drop-in fuel." Yes, give us a fuel that could use the existing distribution system and flow through 190,000 miles of oil pipelines and onwards to 160,000 US service stations. A drop-in fuel could be consumed by existing diesel trucks, locomotives, ships, and diesel-powered machines. Trillions of dollars of infrastructure such as roads, bridges, ports, rail lines, docks, and warehouses have been built to support these diesel workhorses.

Being able to use the existing pipeline system is also the cheapest way to move a fuel. It costs just 1.5–2.5 cents per gallon to move petroleum fuels in a pipe for 1000 miles. Rail costs five times more (7.5–12.5 cents) and trucks 20 times more at 30–40 cents (Curley 2008).

© The Author(s), under exclusive license to Springer Nature Switzerland AG 2021
A. J. Friedemann, *Life after Fossil Fuels*, Lecture Notes in Energy 81, https://doi.org/10.1007/978-3-030-70335-6_6

We are running out of time. Whatever the replacement fuel is, it needs to be commercial before peak oil arrives. After peak oil, it will be increasingly difficult to mount the resources to invent something new quickly enough.

Non-renewable Commercial Liquefied Coal

Liquefied coal is a commercial substitute for diesel that is also a drop-in fuel, though it is not renewable. For the past 50 years, South Africa has liquefied coal for transportation fuels, as did Germany in World War II.

Just five nations have three-quarters of the world's coal. The US has the largest reserves at 24%, followed by Russia at 15%, Australia at 14%, China at 13%, and India 10% (BP 2019).

Höök et al. (2010a) estimated that at best, 54 million barrels per day (Mb/day) of liquefied coal (CTL) could be made worldwide. That is if all coal were devoted to this one goal, and it is still short of the 80 Mb/day of oil produced today. But as little as 17 (Mb/day) of CTL might actually be produced since CTL for transportation will compete with coal required for electricity, cement, steel, iron, and chemicals. The US uses 93% of its coal for electricity production, and 27% of all electricity in the world is generated with coal (EIA 2019, 2020).

Since the thermal efficiency of liquefaction is about 50–60%, only half the coal energy used to make the liquefied coal will come out as energy available in the CTL fuel (Höök et al. 2014). Currently, the US produces 364.5 million metric tons of coal a year. Each ton of coal can make 1 barrel of liquid coal (Höök et al. 2010a). Since at least half of coal energy is lost in making the CTL, the US could make only 182 million barrels of CTL, just 2.5% of the 7.5 billion barrels of oil the US consumes a year (BP 2019). Truly, a drop in the bucket. And it could be even less: Another 40% of the energy dug out of the ground will be lost if carbon capture and sequestration are used, essentially meaning that little or no net fuel would be produced.

It would take decades to double coal production, which requires new coal mines, roads, rail tracks, locomotives, trucks, and water and pipeline infrastructure (GAO 2007; NRC 2009).

The last time the US Geological Survey (USGS) assessed coal resources and reserves was in 1974, and concluded the US had 250 years of coal left. In 2007, the National Research Council wrote a report suggesting 100 years was more likely due to "a combination of increased rates of production, transportation issues, recoverability, and location," and that the USGS ought to resurvey the US to find out (NRC 2007).

The USGS finally reassessed the most important coal basin in 2015, the Powder River basin in Wyoming and Montana, where 45% of US coal is produced. They found there were only about 35 years of coal reserves left, not 250 years (Luppens et al. 2015; Matthew 2016). A 2017 assessment of the Little Snake River and Red Desert coal fields found just 1% of reserves remain (Scott et al. 2017). Many more basins remain to be reassessed, two of them low-energy lignite, perhaps not worth mining economically.

Rutledge (2011) estimated the US had just 63 years of reserves. Today in 2020, if we doubled coal production to make CTL for transportation fuel, then we wouldd have less than 30 years left. But it is questionable whether we can double coal production, since coal production may have already peaked. Heinberg and Fridley (2010) estimate peak US coal occurred in the late 1990s, and Patzek and Croft (2010) say the peak occurred in 2011, based on energy content (since the more energy-dense coal is used first). Reaver and Khare (2014) estimate peak US production between 2009 and 2023.

Importing coal to the US may not be an option either. Globally, peak coal may occur from 2026 to 2034 (Mohr and Evans 2009), 2015–2020 (Zittel et al. 2013), 2025–2030 (Coyne 2016), or 2020–2030 (Höök et al. 2010b).

To many of us, coal is public enemy number one when it comes to the climate. And as for its touted role as an alternative liquefied fuel, it will not take us far down the road. There are however other potential fuels to consider.

Nonrenewable Compressed Natural Gas or Liquefied Natural Gas

Natural gas is finite and so expensive to transport by ship that very little is imported. Over 90% of gas travels in pipelines worldwide.

Natural gas is not a drop-in fuel for diesel engines (though they can be completely overhauled to use natural gas). Natural gas is commercial in that trucks can be built to run on compressed natural gas (CNG) or liquefied natural gas (LNG).

The US has only 6.5% of the world's reserves of natural gas. It is rarely used to fuel vehicles, providing less than 0.1% of US vehicle fuel. The electric grid, manufacturing, and heating consume natural gas, and vehicles have to compete with these existing demands.

Using CNG or LNG as a fuel would require trucks to carry large, heavy cylinders of gas whose high pressure greatly increases the risk of leaks, and requires new or modified engines. And there are few service stations in the US, just 140 LNG stations, each costing up to $2 million dollars, and 1600 CNG stations costing up to $1 million dollars, fueling 160,000 vehicles. About half of CNG is used by private fleets of buses, garbage trucks, and delivery vans, a small fraction of America's 272.5 million vehicles.

Sure, LNG can be imported, but only after many years of building the infrastructure. Massive ships and import terminals must be built, LNG and CNG service stations constructed, LNG trucks, LNG locomotives, and LNG farm equipment created, and natural gas pipelines laid down. Full disclosure: I do not think this is going to happen!

If we are to go down this road, waiting until after peak oil is a bad time to start such a megaproject. That is because oil is the master resource that makes all other goods available, including natural gas and LNG infrastructure.

Peak Natural Gas in the US

In 2005 only about half of America's conventional natural gas remained, depleting at a rate of 5% a year. Then along came fracked natural gas. It too is finite. Fracked shale gas may peak geologically by 2025 and after production peaks, it will decline by 82% within 3 years (IEA 2018, Hughes 2019). Since it often is not profitable—costs exceed revenue for many shale companies, and the coronavirus economy is not helping matters—fracked gas may peak financially even before 2025 (Loder 2015a, b; Rhodes and Hall 2019). McLean (2018) argues that fracked gas might never have happened without historic low-interest rates and Wall Street's willingness to fund these companies. As mentioned in Chap. 2, since 2015, 200 oil companies with over $130 billion in debt have gone bankrupt, and 100 more may fail in 2020.

Despite these business failures, for the past 15 years we have been saved from blackouts and high electric bills by fracked natural gas. And as Hall (2019) points out, prior to the pandemic, cheap fracked oil and gas had kept the economy strong, "subsidized by investors who are either losing money or receiving a poor return on investment. In this respect, (former) President Trump has these financial 'losers' to thank for a large part of the current health of the US economy."

Nonrenewable Noncommercial Exploding Hydrogen

Speaking of great investment scams, I want to say one word to you. Just one word. Are you listening? Hydrogen.

There has to be a lot of whatever it is that replaces fossil fuels. No wonder hydrogen is seen as a possibility since H_2O covers 70% of the planet. Pure hydrogen is the most potent "fossil" fuel (like) substance, so there should be lots of energy there. But oh dear, and sincere apologies for bringing this to your attention, but there are "challenges." Free hydrogen in the atmosphere does not exist, it will escape Earth's gravity or more likely bind tightly with oxygen. And once hydrogen has found a nice oxygen to cozy up to it does not want to leave … no no no. So it takes a whole lot of energy to pry these "abundant" hydrogen atoms away from its buddy oxygen—in fact about 50% more energy than you would gain by reoxidizing it. Foiled again by thermodynamics!

But if you have pure hydrogen (we will not ask where you got it!) AND some pure oxygen you can use them to make an electric current in a fuel cell by reoxidizing them. Although the first hydrogen fuel cell was invented by Sir William Robert

Grove in 1839, hydrogen fuel cells are still far from being commercial for heavy-duty trucks (Friedemann 2016). A big problem is that the specialized membrane required is easily poisoned by impurities in the fuel, and replacing them can be more expensive than the fuel!

Hydrogen is nonrenewable because it is an energy carrier, not a source. It requires nonrenewable natural gas as a source of hydrogen or nonrenewable electric generating contraptions, pipelines, and hydrogen storage tanks that were made using fossil fuels which also provide the energy to split hydrogen from water if coal or natural gas are used to generate the electricity.

Freshwater hydrogen competes with agriculture and drinking water, so ideally hydrogen would be made from abundant seawater. But that requires expensive desalination since electrolysis turns chloride ions into toxic chlorine gas and degrades the equipment. New catalysts, electrode materials, and purification membranes need to be invented (Tong et al. 2020).

The amount of water to make green hydrogen is not a drop in the bucket. After 18 tons of impure water are purified, there will be nine tons of water left, which can be electrolyzed to produce one ton of hydrogen. It will take energy to move all this water to the electrolyzer. Although the electrolyzer facility can be placed next to a river or ocean, this will be possible only if it is a cost-effective place to put a wind or solar farm (Webber 2007; Slav 2020).

Hydrogen is not a drop-in fuel, nor can the natural gas pipeline system or service stations be used for distribution because hydrogen leaks as well as corrodes metal. According to former Secretary of Energy Steven Chu (2020), hydrogen seeps into metal and embrittles it, a material problem that has not been solved for decades and may never be solved.

Meanwhile, hydrogen is stored in expensive austenitic stainless steel containers and pipelines that delay corrosion, and must be carefully maintained and monitored, since embrittlement can result in catastrophic explosions with loss of property and life.

Nor can hydrogen piggyback a ride on natural gas pipelines even at less dangerous concentrations of 5–15%. It can not be extracted at the other end until pressure swing adsorption membranes are invented, and all impurities are taken out so that fuel cells are not damaged. Also, electrolysis of hydrogen from water using excess wind or solar would be able to create hydrogen for only part of the year because both wind and solar are seasonal (Mulder 2014; EIA 2015). Pipeline operators will not want to install and remove expensive hydrogen extraction equipment seasonally (Chatsko 2020).

No container can contain hydrogen for long. Use it or lose it. Hydrogen is the Houdini of elements, the smallest of them all, and will boil off and escape no matter how many gaskets and valves there are on a container and at every pipeline junction.

It is already taken quite a bit of energy for the electrolyzer to split hydrogen from water. To use it for transportation, you will need to do more, which will take more energy than you will ever get back. First, since 99% of hydrogen is made from natural gas rather than water, it will take energy to drill for and distribute the natural gas and second, even more energy to split hydrogen from natural gas by heating it to between 1300 and 2000 °F (700–1100 °C). If you use electrolysis to separate hydrogen from water, that will require even more energy and cost 12 times as much (Romm 2004). At this point, a lot of energy has been used, but the hydrogen still has no usable energy unless you plan to burn it. To store energy for transportation, it needs to be compressed or liquefied to −423 °F using energy-demanding cryogenic support systems that need to run around the clock. Even more energy is required to purify the hydrogen, so fuel cell life and performance are not shortened from clogging with substances like water, carbon monoxide, sulfur, ammonia, and hydrocarbons, which may have been picked up in any of the preceding steps (DOE 2019).

If you wanted to power 5000 hydrogen fuel cell trucks (FCEV) each going 500 miles a day, you would need 10 $75 million dollar hydrogen stations 50 miles apart. That is enough money to build 2500 gas stations. There are no hydrogen stations for nonexistent heavy-duty fuel cell electric vehicles (FCEV). Talk about chicken and egg and the valley of death, who is going to build FCEV trucks for nonexistent $75 million dollar hydrogen stations and vice versa (Zhao et al. 2018).

If the hydrogen is to be delivered, it will take a $250,000 tank truck that can deliver only enough fuel for 60 cars or just a few FCEV trucks. If the tank had gasoline it would have enough for 800 cars. Even more energy is required to keep the tank dispensing hydrogen chilled to −40 °C while fuel cell cars are being filled up, all day long (DOE 2017).

One reason there are no heavy-duty FCEVs is because hydrogen fuel cells last at best 2500 h. Yet to be commercial, they need to last at least 14,560 h. Such a truck would cost $560,000, and a 2500-h fuel cell, which costs $350,000, would need to be replaced six times for $2.1 million additional dollars (den Boer et al. 2013; Zhao et al. 2018; Hunter and Penev 2019).

If heavy cylinders are filled with natural gas at $1 million CNG or $2 million LNG stations and carried onboard, then the truck will operate at only 24.7% efficiency (DOE 2011). Here is the math on that: 84% energy to obtain natural gas * 67% H2 onboard reforming * 54% fuel cell efficiency * 84% electric motor and drivetrain efficiency * 97% aerodynamics, rolling resistance efficiency = 24.7% (DOE 2011). So, somebody explain to me. Why, pray tell, is not the truck just burning natural gas?

If hydrogen escapes it can explode or catch fire ten times more easily than gasoline, set off with just a spark of static electricity. Due to a faulty valve, a Norwegian hydrogen station explosion was so strong that two nearby people inside their cars went to the hospital after their airbags were triggered. Around the same time, a chemical explosion in a hydrogen facility in Santa Clara, California resulted in a cautionary shutdown of all hydrogen stations in the San Francisco Bay Area (Siddiqui 2020). Natural gas has an additive, mercaptan, that imparts a detectable skunky smell but adding an odorant to allow the detection of hydrogen leaks would harm fuel cells. At this point, there are no sensors to detect leaks for pipelines, storage containers, refueling sites, or other enclosed hydrogen storage areas (DOE 2017).

Escaped hydrogen might impact climate change. Hydrogen inevitably escapes during production, transport, or the endpoint of use, which could increase global warming by combining with hydroxyl radicals and reducing their ability to remove methane (Derwent et al. 2020). If hydrogen leakage were as high as 1%, it could have a climate impact of 0.6% that of fossil fuels (Pearman and Prather 2020).

Hydrogen, the next big thing? Okay, looks like I am not cut out to take the job of selling hydrogen futures.

Ammonia and Power-to-Gas (P2G)

Ammonia has been proposed as a fuel, made by splitting hydrogen from water and combining it with nitrogen to make ammonia, also known as fertilizer. Well, good luck, but tick-tock, times a wasting, and this is still a technology in its infancy (Service 2018).

Power-to-gas (P2G) combines hydrogen with carbon dioxide to produce methane. But CO_2 is expensive to capture and there is not enough of it from coal, ethanol, cement, and other industries to produce much fuel, and oil and gas companies will grab what they can to produce more oil and gas. P2G technology is far from commercial, requires reinvention of the Haber-Bosch process, and is unlikely to scale up (Andrews 2015). And, it is not a drop-in fuel.

Noncommercial Oil Shale

How confusing: Oil shale is different from shale oil. The naming committee could have done a better job with this, yes? Shale oil, which accounts for 63% of current US crude production, refers to hydrocarbons that are trapped in formations of shale rock. Oil shale is essentially rock that contains a compound called kerogen.

Despite 90 years of pilot projects, oil shale is still not commercial. The word "oil" is misleading, in that kerogen is a wannabe oil that has millions of years of cooking left to go. Kerogen contains just 10% of the energy of oil, 17% of coal, and 25% of cow dung. Shale oil deposits are mainly in the southwest US, where there is not enough water to process it into a fuel (Udall and Andrews 2005; Cobb 2013).

Pop quiz: Johnny offers you three pounds of kerogen or one pound of cow dung. Which would you choose?

Renewable and Commercial Biodiesel

Thanks for taking the quiz. Now for a more difficult question: What fuels could replace diesel? Looks like our field of candidates with their many false promises has been mercilessly whittled down.

We are down to one candidate: Biodiesel, which is both commercial and renewable. But it is not a drop-in fuel. It can not be used today in most diesel engines at concentrations greater than 20% biodiesel and 80% petroleum diesel. It can not travel in pipelines. But there is so much more to this candidate's story that we will lavish full attention on biodiesel, giving it center stage in later chapters.

But wait—can not we just electrify transportation?

References

Andrews R (2015) Renewable energy storage and power-to-methane. http://euanmearns.com/renewable-energy-storage-and-power-to-methane/. Accessed 1 Nov 2020

BP (2019) BP petroleum statistical review of world energy. British Petroleum

Chatsko M (2020) Does the hydrogen economy have a pipeline problem? https://www.fool.com/investing/2020/08/05/does-the-hydrogen-economy-have-a-pipeline-problem.aspx. Accessed 1 Nov 2020

Chu S (2020) Steven Chu: Lessons from the past and energy storage for deep renewables adoption. Stanford University.https://gef.stanford.edu/events/steven-chu-lessons-past-and-energy-storage-deep-renewables-adoption. Accessed 1 Nov 2020

Cobb K (2013) Geology beats technology: Shell shuts down oil shale pilot project. https://resourceinsights.blogspot.com/2013/09/geology-beats-technology-shell-shuts.html. Accessed 1 Nov 2020

Coyne D (2016) Coal shock model. http://peakoilbarrel.com/coal-shock-model/. Accessed 1 Nov 2020

Curley M (2008) Can ethanol be transported in a multi-product pipeline? Pipeline and Gas Journal 235:34

den Boer E, Aarnink S, Kleiner F, et al (2013) Zero emissions trucks. https://theicct.org/sites/default/files/publications/CE_Delft_4841_Zero_emissions_trucks_Def.pdf. Accessed 1 Nov 2020

Derwent RG, Stevenson DS, Utembe SR et al (2020) Global modelling studies of hydrogen and its isotopomers using STOCHEM-CRI: Likely radiative forcing consequences of a future hydrogen economy. Int J Hydrog Energy 45:9211–9221

DOE (2002) Fuels of the future for cars and trucks. U.S. Department of energy, San Diego, CA

DOE (2011) Advanced technologies for high efficiency clean vehicles. Vehicle Technologies Program. U.S. Department of Energy

DOE (2017) Hydrogen delivery technical team roadmap. U.S. Department of Energy and Partner USDRIVE

DOE (2019) Hydrogen production: Natural Gas reforming. Department of energy, EERE

EIA (2015) Wind generation seasonal patterns vary across the United States. U.S. Energy Information Administration

EIA (2019) Coal explained. Use of coal. Basics. U.S. Energy Information Administration

EIA (2020) Electricity net generation. Total All sectors. Table 7.2a. U.S. Energy Information Administration

Friedemann AJ (2016) When trucks stop running: Energy and the future of transportation, chapter 9. Springer

GAO (2007) Crude oil. Uncertainty about future oil supply makes it important to develop a strategy for addressing a peak and decline in oil production. United States Government Accountability Office

Hall C (2019) Does Trump have a bunch of 'losers' to thank for a growing economy? The Hill. https://thehill.com/opinion/energy-environment/470782-does-trump-have-a-bunch-of-losers-to-thank-for-a-growing-economy. Accessed 4 Dec 2020

Heinberg R, Fridley D (2010) The end of cheap coal. Nature 468:367–369

Höök M, Sivertsson A, Aleklett K (2010a) Validity of the fossil fuel production outlooks in the IPCC emission scenarios. Nat Resour Res 19(2):63–81

Höök M, Zittel W, Schindler J et al (2010b) Global coal production outlooks based on a logistic model. Fuel 89:3546–3558

Höök M, Fantazzini D, Angelantoni A, et al (2014) Hydrocarbon liquefaction: viability as a peak oil mitigation strategy. Philosophical transactions. Series A: Mathematical, Physical, and Engineering Science 372

Hughes D (2019) Shale Reality Check 2019. Post Carbon Institute

Hunter CA, Penev M (2019) Market segmentation analysis of medium and heavy-duty trucks with a fuel cell emphasis. Project ID SA169. U.S. Department of Energy, NREL

IEA (2018) World energy outlook 2018. International Energy Agency

Loder A (2015a) Shale drillers feast on junk debt to stay on treadmill. Bloomberg. https://www.bloomberg.com/news/articles/2014-04-30/shale-drillers-feast-on-junk-debt-to-say-on-treadmill

Loder A (2015b) The Shale industry could be swallowed by its own debt. Bloomberg. https://www.bloomberg.com/news/articles/2015-06-18/next-threat-to-u-s-shale-rising-interest-payments

Luppens JA, Scott DC, Haacke J, et al (2015) Coal geology and assessment of coal resources and reserves in the powder river basin, Wyoming and Montana. United States Geological Service

Matthew B (2016) Amid coal market struggles, less fuel worth mining in US. Associated Press

McLean B (2018) Saudi America: The truth about fracking and how it's changing the world. Columbia global reports

Mohr SH, Evans GM (2009) Forecasting coal production until 2100. Fuel 88:2059–2067

Mulder FM (2014) Implications of diurnal and seasonal variations in renewable energy generation for large scale energy storage Journal of Renewable and Sustainable Energy 6

NRC (2007) Coal. Research and development to support national energy policy. National Academies Press, Washington, DC

NRC (2009) Liquid transportation fuels from coal and biomass: Technological status, costs, and environmental impacts. National Academies Press, Washington, DC

Patzek T, Croft GD (2010) A global coal production forecast with multi-Hubbert cycle analysis. Energy 35:3109–3122

Pearman G, Prather M (2020) Don't rush into a hydrogen economy until we know all the risks to our climate. The Conversation. https://theconversation.com/dont-rush-into-a-hydrogen-economy-until-we-know-all-the-risks-to-our-climate-140433. Accessed 4 Dec 2020

Reaver GF, Khare SV (2014) Imminence of peak in US coal production and overestimation of reserves. Int J Coal Geol 131:90–105

Rhodes C, Hall CAS (2019) The fracking illusion. Royal Society of Chemistry, Chemistry world

Romm JJ (2004) The Hype about hydrogen: Fact and fiction in the race to save the climate. Island Press

Rutledge D (2011) Estimating long-term world coal production with logit and probit transforms. Int J Coal Geol 85:23–33

Scott DC, Shaffer BN, Haacke JE et al. (2017) Assessment of coal resources and reserves in the Little Snake River Coal Field and Red Desert Assessment Area, Greater Green River Basin, Wyoming. Fact Sheet 2019-3053. United States Geological Survey

Service RF (2018) Ammonia-a renewable fuel made from sun, air, and water-could power the globe without carbon. Science

Siddiqui F (2020) The plug-in electric car is having its moment. But despite false starts, Toyota is still trying to make the fuel cell happen. The Washington Post

Slav I (2020) The green hydrogen problem that no one is talking about. Oilprice.com

Tong W, Forster M, Dionigi F et al (2020) Electrolysis of low-grade and saline surface water. Nature Energy

Udall R, Andrews S (2005) The Illusive Bonanza: Oil Shale in Colorado "Pulling the Sword from the Stone". https://www.resilience.org/stories/2005-12-04/illusive-bonanza-oil-shale-colorado/. Accessed 1 Nov 2020

Webber ME (2007) The water intensity of the transitional hydrogen economy. Environ Res Lett

Zhao H, Qian W, Fulton L, et al (2018) A comparison of zero-emission highway trucking technolo-
    gies. University of California. https://doi.org/10.7922/G2FQ9TS7
Zittel W, Zerhusen J, Zerta M (2013) Fossil and nuclear fuels—the supply outlook. https://
    energywatchgroup.org/wp-content/uploads/2018/01/xInnbS-EWG-update2013_
    long_18_03_2013up1.pdfs_.pdf. Accessed 1 Nov 2020

Chapter 7
Why Not Electrify Commercial Transportation with Batteries?

When It Comes to Diesel, Electric Cars Are Irrelevant

In our quest to replace diesel, electric cars are of no help because cars run on gasoline. Let me explain.

A barrel of oil is refined into many products—gasoline, diesel fuel, heating oil, jet fuel, petrochemical feedstocks, waxes, lubricating oils, and asphalt. Only so much of each product can be produced from a barrel of oil. The refinery operator can not just say hey, today is a fine day to make diesel. Sad to report, despite the alchemical wonders of a refinery, there are limits on how much diesel can be extracted from a barrel of oil. The limit is about 28% of the barrel. Gasoline can comprise about 40% of a barrel. That gasoline fraction of the barrel can not be made into diesel to keep trucks, locomotives, or ships running. US diesel petroleum refineries produce about 10–12 gallons of diesel out of a 42-gallon barrel of oil.

Like I said, electric cars do not solve our diesel problem.

But what about battery-powered trucks?

The Heavy Lift to Improve Transportation Batteries

Batteries and fuel cells have been around since 1800. Ever since then people have hoped a battery powerful enough for transportation would be invented (Hiscox 1900). But the energy density of batteries has increased only sixfold since the first lead-nickel batteries debuted a century ago (Van Noorden 2014).

To make a better battery, consider what it takes: (1) high energy density for hours of power and range; (2) high power density to accelerate; (3) rechargeable thousands of times; (4) retains 80% of energy storage capacity for 10–15 years; (5) fast recharge; (6) reliable; (7) not harmed by over or undercharging; (8) able to withstand

A. J. Friedemann, *Life after Fossil Fuels*, Lecture Notes in Energy 81, https://doi.org/10.1007/978-3-030-70335-6_7

the vibration and shaking of rough roads; (9) made out of cheap, common, nontoxic, and recyclable materials; (10) will not catch fire or explode; (11) does not discharge while not in use; (12) performs well in cold and hot temperatures; (13) low recycling cost; (14) and little maintenance. Wait, there is more: To enable longer range, the battery must be as small and lightweight as possible. Whew!

Every time a battery is tweaked to improve just one of these qualities, others might be diminished or lost. Each improvement needs to be tested for the equivalent of 10 years of battery cycles, the minimal lifespan of a vehicle battery. Testing and waiting is slow and it is expensive to hire material scientists, chemical, mechanical and computing engineers, electrochemists, nanotechnologists, and more (Borenstein 2013). The battery may be a simple black box, but what goes on inside is a complicated choreography of electrons.

Batteries can not be made lighter, because they are already made out of lithium, the third lightest element. Forget about the two lighter elements, there will never be ghostly hydrogen or helium gas batteries. Lithium is finite, there is not enough of it to switch our entire fleet to electric cars let alone electric trucks and utility-scale energy storage for the electric grid (Vazquez et al. 2010; Vikström et al. 2013). Other minerals used in lithium-ion batteries are finite too, like cobalt (Patel 2018).

Further limiting possibilities of lightening batteries for trucks is that the palette of elements in the periodic table is limited to 118, far less than the hundreds of paint colors at art supply stores. Most elements can be ruled out: 39 are radioactive, 23 too scarce or expensive, 6 noble gases are inert, while others are too heavy, too hard to recycle, or too toxic (cadmium, cobalt, arsenic, etc.).

Most elements do not have enough reduction or oxidation potential to generate power. The elements that could produce the most are the strongest reducing element lithium and the strongest oxidizing element fluorine, but after 60 years of research, this lithium fluorine battery may never come to fruition (Scrosati et al. 2013).

Diesel is the Mr. Universe of fuels. Because of the laws of physics and thermodynamics and what we know from two centuries of research, it is not likely that any kind of battery could ever approach the energy density of diesel fuel, which at 13,762 Wh/kg is 61–117 times more powerful than a lithium-ion battery at 118–225 Wh/kg (Cushman-Roisin and Cremonini 2020).

Batteries Have a Weight Problem

The Tesla Model 3 battery pack weighs 1054 pounds (480 kg), over a quarter of the total weight of the car. Because of battery weight, heavy-duty trucks (class 7 and above) are too heavy to carry much cargo. Plus moving the extra battery weight takes too much energy. There are medium-sized delivery vans that run on batteries, dependent on braking often for recharge, then fully recharging overnight locally. These trucks are heavily subsidized by state clean air programs since they cost substantially more than a diesel truck. California has less than a thousand electric trucks out of the 1.5 million trucks operating in the state (O'Dea 2019).

Battery-powered commercial heavy-duty trucks—class 8 trucks, but you call them semitrucks or 18-wheelers—do not exist because they are far too heavy and expensive and can carry very little cargo. An empty class 8 *diesel* truck weighs about 35,000 pounds and can haul 45,000 pounds of goods before hitting the US maximum road weight of 80,000 pounds.

How much cargo can a battery-powered electric class 8 semitruck haul? Different researchers cited below reached the following conclusions:

- A truck with a driving range of 600 miles needs a battery pack weighing 35,275 pounds and can carry just 10,000 pounds. For 900 miles, the truck needs a 54,000-pound battery and even with no cargo would weigh 89,000 pounds, exceeding the maximum road weight (Sripad and Viswanathan 2017).
- A study by den Boer et al. (2013) reached a similar conclusion. A 621-mile trip would require a battery weighing as much as 55,116 pounds.

Another weighty issue is charging time. A truck capable of going 100 miles—it would need a 350-kWh battery—takes over 12 h to recharge. Ultra-fast charging is not an option since it can reduce battery lifespan, out of the question for such an expensive battery. The port of Los Angeles ruled this kind of truck out for their clean air program because the battery weight cut too much into the cargo weight that could be hauled (Calstart 2013; Sripad and Viswanathan 2017).

Another timely issue is the expense. To go 500 miles requires a 30,000-pound battery that would cost you $350,000, plus another $100,000 for the truck itself (Calstart 2013).

And forget about batteries to power ships. On a container ship carrying 18,000 20-foot-long containers on a month voyage from Asia to Europe, you would need 100,000 metric tons of batteries taking 40% of the cargo space (Smil 2019).

There is no polite way to say this: Batteries have a weight problem. And there is no ready remedy. Batteries are housed within a battery management system (BMS), a steel case to protect the vehicle from fires (lithium is quite flammable) plus monitoring devices that reduce battery efficiency and cooling systems.

Fast Charging

Wanted: Truck drivers with the patience of Job, willing to meditate over long hours several times a day while their batteries are recharged. Benefits include frequent bathroom stops… Pull that help wanted ad down. Looks like we have no takers.

Failing to find Zen truck drivers, we could just speed up the charging process. Call it fast charging. The only problem is this is not yet possible. Oxford professors estimated that the power needed to charge just one truck's battery using fast charging in 30 min would use, over the course of a year, as much power as 4000 households. Such fast charging is not possible yet and would put the electric grid under enormous stress (Harris 2017).

But fast charging trucks is essential. Truckers can not sit around for 12 unpaid hours honing life skills and learning to crochet while waiting for the battery to recharge. But fast charging trucks may never be possible. Scientists at U.C. Riverside recently fast charged batteries similar to Tesla batteries using existing highway fast charging technology. They found that batteries cracked, leaked, lost storage capacity, and suffered internal chemical and mechanical damage, reducing their lifespan. The high heat generated is also a danger that could lead to fire or explosion in the 7104 lithium-ion batteries in a Tesla Model S or the 4416 in a Tesla Model 3 (Quimby 2020).

But What about the Tesla Semitruck?

You can reserve yours now with a $20,000 deposit. When you will get your truck … nobody knows. If the Tesla truck ever does appear, it might take 5 years or more to find out if the battery life and driving range was reduced from fast charging, which Tesla claims can be done in 30 min. On top of that, it remains to be seen how much other factors affect battery life, such as the thousands of miles of rough rides trucks endure, and cold weather, which reduces range up to 35% (Sharpe 2019).

Off-Road Trucks Are Off-the-Grid

Looking farther out, how will off-road trucks (e.g., military, mining, logging, backhoe, road construction, and maintenance) that work far from the grid recharge? I do not think they will. Nor is there enough power in rural areas to charge thousands of agricultural tractors, combines, and other equipment all at the same time during the short planting or harvest windows. Heavy batteries weigh enough to compact soil and reduce future harvests. With their weight problem, electric tractors very well could end up stuck in the mud.

References

Borenstein S (2013) What holds energy tech back? The infernal battery. Associated Press. https://news.yahoo.com/holds-energy-tech-back-infernal-battery-221338549%2D%2Dfinance.html
Calstart (2013) I-710 project zero-emission truck commercialization study. Calstart for Los Angeles County Metropolitan Transportation Authority. 4.7. https://calstart.org/libraries-i-710_project-i-710_project_zero-emission_truck_commercialization_study_final_report-sflb-ashx/
Cushman-Roisin B, Cremonini BT (2020) Useful numbers for environmental studies and meaningful comparisons, chapter 3. Dartmouth college under contract with Elsevier https://www.dartmouth.edu/~cushman/books/Numbers.html

den Boer E, Aarnink S, Kleiner F et al (2013) Zero emissions trucks. In: An overview of state-of-the-art technologies and their potential. CE Delft, Delft

Harris B (2017) Tesla's electric truck 'needs the energy of 4,000 homes to recharge', say researchers. World Economic Forum. https://www.weforum.org/agenda/2017/12/tesla-s-electric-truck-needs-the-energy-of-4-000-homes-to-recharge-say-researchers/. Accessed 2 Nov 2020

Hiscox GD (1900) Horseless vehicles, automobiles, motor cycles operated by steam, hydrocarbon, electric and pneumatic motors. Norman Henley & Co.

O'Dea J (2019) How can we get more electric trucks on the road? Union of Concerned Scientists. https://blog.ucsusa.org/jimmy-odea/how-can-we-get-more-electric-trucks-on-the-road. Accessed 2 Nov 2020

Patel P (2018) Could Cobalt choke our vehicle future? Demand for the metal, which is critical to EV batteries, could soon outstrip supply. Sci Am

Quimby T (2020) Debate over DC fast charging points to fleet needs, expectations. https://www.ccjdigital.com/dc-fast-charging-points-fleets/. Accessed 2 Nov 2020

Scrosati B, Abraham KM, von Schalkwijk W et al (2013) Lithium batteries. Advanced technologies and applications. Wiley & Sons, Hoboken, NJ

Sharpe B (2019) Zero-emission tractor-trailers in Canada. International Council on Clean Transportation

Smil V. 2019. Electric container ships are stuck on the horizon. Batteries still can't scale up to power the world's biggest vessels. IEEE spectrum. https://spectrum.ieee.org/transportation/marine/electric-container-ships-are-stuck-on-the-horizon. Accessed 2 Nov 2020

Sripad S, Viswanathan V (2017) Performance metrics required of next-generation batteries to make a practical electric semi-truck. ACS Energy Letters 2:1669–1673. https://doi.org/10.1021/acsenergylett.7b00432

Van Noorden R (2014) The rechargeable revolution: a better battery. Nature 507:26–28

Vazquez S, Lukic SM, Galvan E et al (2010) Energy storage systems for transport and grid applications. IEEE Trans Ind Electron 57:3881–3895. https://doi.org/10.1109/TIE.2010.2076414

Vikström H, Davidsson S, Höök M (2013) Lithium availability and future production outlooks. Appl Energy 110:252–266

Chapter 8
Catenary Electric Trucks Running on Overhead Wires

Overhead electric catenary wires that power trolleybuses and trams have existed since the 1880s. Today, San Francisco has 273 trolleybuses and 200 tramcars riding the wires. For carrying passengers, they work well. Plus, they are charming.

In Southern California, the ports of Los Angeles and San Pedro would like to replace over 16,000 diesel drayage (short-haul) trucks that carry containers from ships to warehouses 25 miles or more inland on I-710 with new trucks that would use an electric catenary truck route to reach zero emissions (Calstart 2013).

Will It Work?

The ports are talking about a fleet of 16,000 trucks riding the overhead wires. These 16,000 drayage trucks would be twice as heavy as San Francisco's trolleybuses and would roll down the I-710 highway practically on top of each other, seconds apart, not spaced out several minutes apart like trolleybuses.

No one knows if a catenary system can handle so many trucks at once or how many new power plants would be required. But there are three projects exploring the possibility in Southern California, Germany, and Sweden with one to five-mile-long catenary systems. Since all of these projects are testing with just a few trucks, we still do not know how many trucks such a system could support (Siemens 2018; Zhao et al. 2018).

A. J. Friedemann, *Life after Fossil Fuels*, Lecture Notes in Energy 81, https://doi.org/10.1007/978-3-030-70335-6_8

A Catenary System Would Cost a Truckload of Money

A catenary system of support structures, lines, insulators, power-control systems, power lines, and substations costs from $5–$7 million to $9–$12.8 million per mile (Siemens 2018). Stringing up the 25 miles of the I-710 corridor would cost $125 to $320 million. Building an extensive catenary system for just the 7626 miles of major federal and state highways in California (FWHA 2001), if it were possible, would cost many truckloads of money. And billions more for additional electricity generation for the million trucks that travel 25 billion miles a year on California roads (Zhao et al. 2018). Trucks entering California from other states presumably would not be adapted for catenary wires nor would California catenary trucks be able to traverse long distances outside the state on their secondary propulsion system.

Then add another $90,250 per mile of catenary system for maintenance (Siemens 2018 page 82). The wires need to be 18 ft above the ground, so there will be many obstacles such as road overpasses, trees, and traffic lights, adding to the cost of installation. Overhead lines swing in high winds, lose power when struck by lightning, and are coated with ice that can cause poor connections, electrical arcing, and power surges.

To ride the wire, catenary vehicles rely on a doohickey mounted on the roof to collect power through contact with the overhead line. Okay, I call it a doohickey. They call it a pantograph. Catenary trucks differ from trains, catenary buses, and trams because they need a new type of pantograph to connect and disconnect at high speeds from the overhead wires. Catenary trucks also require a second method of propulsion for times when they leave the wire to pick up and deliver cargo, pass other trucks, or get around obstacles.

Dual Propulsion Doubles to Triples the Cost

The second source of power for a catenary truck would need to be a hydrogen fuel cell (FCEV) or a battery (BEV) truck, neither of them commercial. Diesel would be an option until it becomes unavailable from depletion or being rationed to agriculture and other essential services.

A new drayage diesel truck costs $100,000, plus a $50,000 pantograph to connect to the wires. I have to respect that anything costing $50,000 is not a doohickey.

Where was I? Oh yes, having two modes of operation doubles or triples the cost. A battery truck (BEV) will cost at least $300,000 ($100,000 truck + $105,000,350 kW battery + $50,000 pantograph + $33,000 chargers).

A fleet of 15,000 BEVs will also need $500 million in battery chargers for the 24 miles of the I-710 corridor using 500 fast charger units and 7500 overnight slow chargers. Please note that will require 47 acres of land (Siemens 2018).

A hydrogen FCEV truck will cost $560,000, including the truck and battery (DOE 2019) plus $50,000 for the pantograph, and $750 million for ten hydrogen stations, each 50 miles apart on a 500-mile roadway (Zhao et al. 2018; Hunter and Penev 2019).

Holey moly! Given what is involved, it is no wonder that the technology for truck catenary, battery electric drayage, and hydrogen fuel cell trucks does not exist. Nor do the dozens of new electrical generation plants or hydrogen service stations to power them.

Even if the money could be found, then what? Catenary wires are 18-feet high, so residential roads could not be wired for catenary trucks; there are too many trees, utility poles, and lines in the way.

Nor can trucks drive from city to city. Freight traveling 2000 miles on the wire would require at least three new large power plants of 500 MW each (FRA 2009) and 2000 miles of catenary overhead lines plus dozens of battery charging depots or hydrogen fuel stations.

Conclusion

It may not be possible to run thousands of trucks on a catenary system. It is never been done before. With such high costs, catenary systems would be out of the question other than for sections of roads with very high truck traffic. Off-road trucks will never run on the catenary, especially the most important trucks: Agricultural tractors work 360 million acres of farmland. Off-road trucks are indispensable for mining, logging, and maintaining transmission lines, pipelines, railroads, and dams on 1.35 million miles of unpaved roads. No farmer, no miner, no logger, and very few truckers will ever go catenary.

References

Calstart (2013) I-710 project zero-emission truck commercialization study. Calstart for Los Angeles County Metropolitan Transportation Authority. 4.7. https://calstart.org/libraries-i-710_project-i-710_project_zero-emission_truck_commercialization_study_final_report-sflb-ashx/

DOE (2019) Market segmentation analysis of medium and heavy-duty trucks with a fuel cell emphasis. Department of Energy, National Renewable Energy Laboratory

FRA (2009) Federal railroad administration comparative evaluation of rail and truck fuel efficiency on competitive corridors. ICF International for U.S. Department of Transportation, Washington, DC

FWHA (2001) National highway system length 2001. Miles by functional system. U.S. Department of transportation, Federal Highway Administration

Hunter CA, Penev M (2019) Market segmentation analysis of medium and heavy-duty trucks with a fuel cell emphasis. Project ID SA169. U.S. Department of Energy, NREL

Siemens (2018) Construction of a 1-mile catenary system and develop & demonstrate catenary hybrid electric trucks. eHighway SoCal Final report. SCAQMD contract 14062. Siemens Industry INC.

Zhao H, Qian W, Fulton L, et al (2018) A comparison of zero-emission highway trucking technologies. University of California. https://doi.org/10.7922/G2FQ9TS7

Chapter 9
Manufacturing Uses Over Half of All Fossil Energy

Cars, trucks, trains, ships, our homes—even taken together, do not account for the majority of our energy use. The biggest energy glutton is manufacturing, the Godzilla of energy users. The industrial sector consumes 54% of all world energy and 35% of all US energy (EIA 2019a, b).

Globally, 73% of all coal is used by industry, 37% of natural gas, 7.2% of oil, and 42% of electricity. The US manufactures 18% of the world's goods (West and Lansang 2018) with an energy mix of 40% natural gas, 34% oil, 12.3% electricity, 9% biomass residues, and 4% coal in the industries shown in Fig. 9.1.

Industry's energy needs are different by degree from those required by transportation and buildings. Industry often relies on high heat. Just two examples: Absent high heat, we would not have steel and cement. Industry does not draw most of its energy from the electrical grid. Instead, it uses onsite kilns, furnaces, and boilers, all fired by fossil fuels, to produce high heat. In the US about 75% of industrial energy is used to generate heat, with 83% of the heat generated by fossil fuels. Heat can be used directly, as steam heat, or to generate reliable electricity at the factory to operate machine drive equipment (DOE 2014).

As you can see in Fig. 9.2, which represents all US manufacturing, 27% of the fossil energy is lost off-site (14,759 output/20,008 input) before reaching the factory. Another 50% of the energy is lost onsite (7427/14,759) at the factory in electricity and steam generation losses, steam distribution losses, process losses, and non-process losses (Brueske et al. 2012).

Only 23% of the energy is converted into usable work. Research to improve efficiency would reduce fossil use and CO_2 emissions to buy time to transition to something else (DOE 2004).

Little known industrial-strength fact: Most products have no known way of being made with electricity or renewables. Nor is there much interest in doing so, with very little R&D being done. Even if electrical, hydrogen, or other alternative means

A. J. Friedemann, *Life after Fossil Fuels*, Lecture Notes in Energy 81, https://doi.org/10.1007/978-3-030-70335-6_9

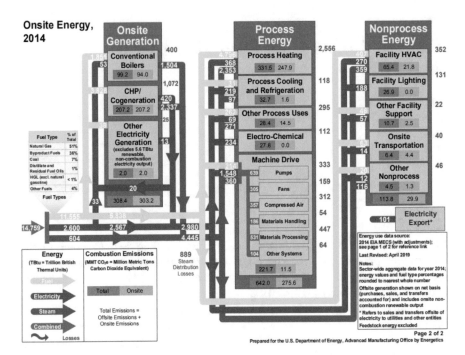

Fig. 9.1 Onsite energy use of all manufacturing in the US, combines the footprints of 94% of manufacturing energy used for: Alumina and aluminum, cement, chemicals, computers, electronics, electrical equipment, fabricated metals, food and beverage, forest products, foundries, glass, iron and steel, machinery, petroleum refining, plastics, textiles, transportation equipment. (Sources: Brueske et al. 2012; DOE 2014)

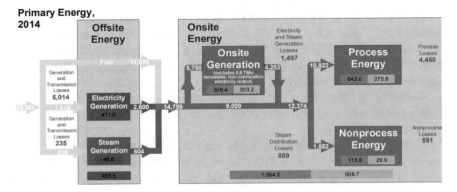

Fig. 9.2 Overview of all manufacturing energy in the United States: Fuel, electricity, and steam generation. (Sources: Brueske et al. 2012; DOE 2014)

of production were made commercial, there are formidable barriers to making a transition. Industrial profit margins are razor thin and companies compete internationally. Existing blast furnaces, kilns, catalytic crackers, and other fossil-fueled components can cost hundreds of millions of dollars, often existing within multibillion-dollar facilities that have 30–60-year lifetimes (Friedmann et al. 2019; Sandalow et al. 2019). It would cost a fortune to modify or replace them.

If manufacturers were forced to decarbonize with alternative energy to comply with greenhouse gas regulations and the costs were high, facilities are likely to be offshored to nations without regulations that still had oil, coal, or gas, taking jobs and revenue along with them, increasing prices and unemployment at home. That is a problem.

Governments are not likely to require alternative energy retrofits since many industries are politically essential. No nation wants their auto, airplane, cement, steel, and petrochemical factories to go offshore, which is why they are often exempt from environmental regulations and carbon limits (Friedmann et al. 2019).

Completely new equipment and processes need to be invented to replace fossil fuels for nearly all kinds of industry. There are no one-size-fits-all solutions. Facilities vary. For example, the chemical industry has complex facilities with dozens to hundreds of burners, furnaces, and boilers, many of them fit-to-purpose and not readily replaced (Malico et al. 2019; Sandalow et al. 2019).

Renewable energy solutions need to be local, since one or more of biomass, wind, solar, nuclear, energy storage, and a much larger transmission system may not be available for a given manufacturing facility (Friedmann et al. 2019).

Another manufacturing issue confronting us is that half a million products, such as plastics and paint, are made out of fossil fuels and will need to be made from something else (Sandalow et al. 2019).

Industrial Fossil-Fueled High Heat Makes Cement, Steel, Roads, Bridges, Dams, and Buildings

As we struggle to replace fossils with renewables, one key element that fossils bring to the table is consistently ignored or overlooked: Its unmatched ability to generate high heat. Industry requires very high heat to forge concrete and steel, the backbone of civilization. Today, this high heat is generated by burning fossil fuels. High heat and fossil fuels also are essential for creating millions of other products such as fertilizers, glass, plastics, rubber, ceramics, computers, chemicals, and tools.

Even if there were electrical processes to replace fossil fuels in industry, the current US electric grid is badly mismatched for the job. Why is that? Because two-thirds of US electricity comes from fossil fuels, and 66% of coal and natural gas energy is lost as heat at the power plant when making electricity. Yet another 4–10% is lost over the wires, so only 24–30% of the energy contained in fossil fuels reaches a factory. Right now, it is three times more energy-efficient to create heat directly

Table 9.1 Manufacturing temperatures, energy, operations, and applications

Fahrenheit	%	Manufacturing operation	Application Examples
1710–3000	3.7	Nonmetal melting	Plastics, rubber, food preparation, softening
1330–3000	17.8	Smelting and metal melting	Steelmaking and other metal production, glass, ceramics
1150–2640	7.3	Cement	Calcining 1652 F Sintering 2640 F
1330–3000	3.7	Metal heat treating and reheating	Hardening; annealing; tempering; forging; rolling
710–2010	1.7	Coking	Ironmaking and other metal production
320–1020	21.6	Drying	Water and organic compound removal
280–1200	2.0	Curing and forming	Coating; polymers, enameling, molding, extrusion
230–860	29.3	Fluid heating	Food prep; chemicals; distillation; cracking
210–3000	12.8	Other	Incineration; preheating; catalysis

Source: DOE (2015a, b), Friedmann et al. (2019), Sandalow et al. (2019)

through the burning of fossil fuels than with electricity (EIA 2013). More wind and solar electricity would narrow that efficiency gap, but not erase it. Currently, only 9% of electricity comes from wind and solar (EIA 2020a).

The idea that our energy problems are solved when renewables can generate electricity for the same price as fossil fuels—so-called grid parity—is not true. To replace fossil fuels, renewables must also reach "thermal parity" by powering new processes that can produce the high heat of up to 3000 F needed for the essential manufacturing processes listed in Tables 9.1 and 9.2.

Renewable High Heat must be Reliable

Most iron and steel are made in enormous blast furnaces that can run continuously for up to 20 years. Unexpected outages can damage the brickwork lining.

Computer chip fabrication plants need to run continuously for weeks to accomplish the thousands of steps needed to make microchips. A half-hour power outage at Samsung's Pyeongtaek chip plant caused losses of over $43 million dollars (Reuters 2019a).

Unexpected outages can leave materials cooling in tanks and pipes, causing them to crystallize or harden, clogging the pipes. Many processes need an exact continuous temperature and pressure because variations can cause metal fatigue and wear and tear.

Even facilities that do not run continuously need to be up 60–95% of the time to repay their high capital investment.

A renewable industrial energy solution must be able to generate high heat from a few hours to around the clock.

Table 9.2 Maximum heat generated by nonfossil energy sources

Heat source	Maximum temperature fahrenheit	Comment
Biomass: Fuel	4000	Biodiesel, ethanol
Hydrogen (H$_2$)	3800	Made from natural gas or electrolysis of water
Electric: Resistance	3275	Indirect heat
Solar: Parabolic dish	2200	Small area heated
Biomass: Charcoal	2010	From forests, agriculture, waste
Concentrated Solar Power (CSP)	1800	Small area heated
CSP oven	1800	Not commercial, small area heated
Nuclear: Advanced	1562	Not commercial
Biomass: Birch wood	1742	Depends on the tree, i.e., redwood is 690 F/364 C
Molten Salt	1040	Thermal Energy storage
Solar: Parabolic trough	752	Same for solar linear fresnel
Nuclear: Conventional	572	Third-generation reactors
Geothermal	380	
Electric: Microwave direct heat		Temperature depends on material

Source: DOE (2015a, b), Friedmann et al. (2019), Sandalow et al. (2019)

For steel and cement, alternative approaches might include producing in batches rather than continuously. But even if this were possible, that would be less energy efficient, cost more, and produce far less (Heinberg and Fridley 2016), with a risk of bankruptcy or offshoring.

Microchips, some chemicals, and other products might not be possible to produce in batch mode.

There Is No Way to Store High Heat

What happens when the wind dies down or the sunsets? An electric grid mainly dependent on wind and solar will need to have expensive backup electricity storage in order for industry to rely on it. In addition, enormous amounts of high heat will need to be stored to keep manufacturing processes running. Currently, there are no means to store hours of high heat. If there were, factories would need to be within a few hundred feet of the storage or source since the heat cools off rapidly (Oberhaus 2019).

Electrifying Manufacturing

Inventors, have I got a job for you: Imagine, research, develop, and implement new ways of producing goods with electricity that are now made with fossil fuels. Required skills: A tolerance for high heat. Must not be easily discouraged. Lots of work here!

One notable advance in electrifying manufacturing involves scrap steel. Existing scrap steel can be recycled using electricity to power an electric arc furnace, which uses high-current electric arcs. A quarter of all steel is made that way. The rest is made from scratch for a number of reasons, including a lack of scrap steel or enough electricity. An electric arc furnace can use as much electricity as a town of 100,000 people (Brown 2016). Additionally, every time steel is recycled it is degraded to a lower quality than the original steel, which was often custom-made with alloys (i.e., chromium, cobalt, etc.) for a particular application—to be hard, strong, durable, or corrosion-resistant. Because of this, recycled steel can not be used in many industries (Bardi 2014).

Got an idea for how to make cement using electricity? Send it my way. We will be rich! Because right now, there is no known way to use electricity to do this. Cement is made in 131-foot-long (40 m) kilns in two stages, calcining at 1832 °F (1000 °C) and clinkering at 2550–2730 °F (1400–1500 °C). Kilns can run continuously for 18 months. It is hard to imagine how that could be done with erratic renewable electricity and no high-heat storage. Cement kilns are heated internally with fossil fuels. External resistance electric heat can not heat the kiln uniformly, and resistance heating technology is still immature (Friedmann et al. 2019).

All contraptions that produce electricity need high heat in their construction. They all need cement made at 2600 °F. Those made with steel, like wind turbines, need temperatures of 3100 °F (1700 °C) (WCA 2020). Solar panels require 2700°–3600 °F (1500–2000 °C) heat to transform silicon dioxide into metallurgical grade silicon (Honsberg and Bowden 2019).

Manufacturers Have to Move Next to a Thermal Heat Source If It Can Not Be Stored

Here are the possible alternative sources of thermal heat to replace fossil heat:

Nuclear Power. Industries that do not need very high heat could try to obtain thermal heat near one of the 58 remaining nuclear plants in the US if there is real estate to do so. But these generate a maximum of 572 °F (300 °C), and many of them are likely to close in the future (Cooper 2013). New generation IV nuclear plants are on the drawing boards but do not exist yet, and can generate only 1562 °F (850 °C) heat, not enough to make cement, steel, glass, and more.

Geothermal Power. There are only six plants that generate 50 MW or more with a maximum heat of 380 °F (193 °C), too low for most industrial needs.

Solar Collectors can generate heat up to 400 °F (200 °C) (EPA 2017). Large-scale parabolic trough collectors can generate heat as high as 752 °F (400 °C) (Moya 2012). While not enough heat for many industrial processes, it is enough to make beer, which perhaps began civilization (Kahn 2013). So for sure add "build solar collectors" to the list of what we need to do after you finish reading this book.

Concentrated Solar Powered (CSP) plants use mirrors or lenses to focus and concentrate a large area of sunlight onto a receiver and produce heat and steam to drive a turbine. CSPs are designed to generate electricity, not the high heat of manufacturing. In today's solar thermal plants, solar energy is converted into steam (via a steam boiler), which is then converted into electricity (via a steam turbine that drives an electric generator). Sad to say, this process is just as inefficient as converting electricity into heat: Two-thirds of energy gets lost when converted from steam to electricity.

Very few CSP plants have molten salt, which is used to store energy to generate electricity for a few hours after the sun goes down. If this heat were instead used for manufacturing, at best its maximum working temperature is 1040 °F (560 °C) (Friedmann et al. 2019).

There is a small, experimental CSP oven made by Heliogen that can generate 1800 °F (1000 °C) using 400 mirrors the size of a large flat-screen TV on a third of an acre aimed at a small square-foot target area on top of a 55-foot tower. Perhaps a dozen kilograms of cement a day could be made in this small space. As they say, Rome wasn't built in a day!

To make cement at a commercial scale of several thousand tons a day would require a Heliogen system the size of the 5.4 square mile billion-dollar Ivanpah CSP plant in the Mojave Desert (NREL 2013). The focal hotspot would be 460 feet in the air. Since heat is generated on only one side, not useful for most industrial processes, the solar heat will need to warm up some other medium that can transmit the heat downward and inward to a facility, losing some of its temperature along the way (Barnard 2019; Oberhaus 2019).

Even so, it is not clear how much cement could be made since there are only a few hours of the day with maximum heat generation from the sun. The average capacity factor for the past 10 years for CSP is quite low, only 22% in the US from 2010 through 2019, so no power is generated, on average, 78% of the time (EIA 2020b).

Nearly all CSP plants in the world exist in deserts, where humidity does not reduce solar radiation. In the US, seven of the eight solar thermal power stations are in the southwest. US industry consumes a huge amount of water, 14.8 billion gallons a day (Dieter et al. 2015), so few industrial facilities could take advantage of CSP-generated heat.

Hydrogen

Steelmaker ArcelorMittal says that carbon dioxide-neutral steel production using hydrogen is *theoretically* possible. Apparently not enthusiastic about the prospect, the company (and I am paraphrasing here) says that you would have to pry their fossil-fueled blast furnace out of their cold dead hands. The company went on to say that commercialization of hydrogen steel-making is decades away and would drive steel costs up 60–90%. ArcelorMittal says it would cost up to $786 billion to switch from coal to hydrogen made with water electrolysis using wind and solar power, and take decades to scale wind and solar up enough to produce sufficient power to create the hydrogen and energy for the steel industry to use. If biomass and plastic waste were used to make steel, ArcelorMittal says it would cost $168–$246 billion to sequester 150–200 million tons of CO_2 a year. A thousand dollars a ton! Nor is it technically possible to make a complete shift to bioenergy (AM 2020; McDonnell 2020; Pooler 2019). Anyone uncertain on where the company stands on this?

With half of the world production coming from coal-rich China, competing on price is not negotiable. Using natural gas as a feedstock is the preferred way to make hydrogen and today that is how over 96% is made, emitting 830 million tons of CO_2 a year. Only 1% of hydrogen is made from water with renewable electricity (Frangoul 2020).

To make steel, an immense amount of hydrogen would be required. It takes 780,000 metric tons of coal to make a billion tons of steel.

So steel is a heavy lift. What about hydrogen-fired cement? Sad to report, as with steel, cement also depends on coke (a fuel made by heating coal or oil in the absence of air). There is no process to make cement using hydrogen. Even if a way could be found to burn hydrogen instead of coke to make cement, the clinker burning process would have to be significantly modified, because of hydrogen's low radiation heat transfer. Hydrogen flames are not suitable for clinker burning.

Full disclosure: I should mention that hydrogen is explosive!

Most industries would be reluctant to use hydrogen because it is explosive and difficult to use (Heinberg and Fridley 2016; Friedmann et al. 2019). Hydrogen always leaks (it is the smallest element), so eventually a factory risks being blown up, since it is easy to ignite, 12 times easier than gasoline vapor. The smallest of sparks can turn hydrogen into a bomb (Huang 2019; Reuters 2019b; Szymkowski 2019). Remember the Hindenburg!

If hydrogen were made off-site to reduce explosion risks, the pipelines, storage tanks, compressors, and chillers would cost 10 times more than similar natural gas infrastructure, because expensive metal immune to hydrogen embrittlement and to prevent its escape is required (ECRA 2007). A distribution infrastructure similar to natural gas, with 300,000 miles of long-distance pipe and 2.1 million miles of distribution pipelines would need to be built. But there are only 620 miles of hydrogen pipelines in the US, mainly along the Gulf Coast for oil refining and ammonia fertilizer production.

Power2gas Aka Power-to-Methane

Power2Gas is a technology that uses electrical power to produce a gaseous fuel, such as methane or hydrogen. Power2Gas is far from commercial. This technology exists only at a laboratory scale, is very expensive even with low-cost renewable power, does not have enough CO_2 or biomethane sources to scale up, and has a very low round-trip efficiency of 20–30%, which means 70–80% of energy input is lost (Palmer 2020; Friedmann et al. 2019; SBC 2014; Andrews 2015).

Power2gas—please do not ever mention it again!

That Leaves Biomass, Once Again, as Our Post-carbon Savior

We know charcoal from biomass can make iron and steel because that is how it was done before fossil fuels. Even today, Brazil uses biomass, producing 5.2 million metric tons of steel in 2017 with charcoal in small kilns, at great ecological cost (Munnion 2020).

But it will not be easy. It is difficult to conceive of hundreds of bonfires burning around the clock in refineries, chemical and other continuous, high-capacity processing facilities. And already forests are being way overused.

Many industrial burners are fed by pipeline with natural gas or oil. Although biodiesel could be used, most if not all biodiesel likely will be used for transportation to bring in the materials required for manufacturing.

Instead, the high heat would be generated by first converting wood to charcoal. But even that is temporary. Biomass supplies are economic within 50 miles of a factory or electric generation facility today due to plentiful and cheap diesel. The range will shrink when diesel is scarce or wildfires increase from climate change. After deforestation, the facility will go out of business for 50–100 years until a new forest grows.

Biogas—fuel that is naturally produced from the decomposition of organic waste—does not scale up since its source is landfills that are unlikely to be near manufacturing and that produce only miniscule amount of gas.

Brazil does use wood to make steel. But at a steep price. Half of Brazil's wood comes from illegally harvesting native forests, resulting in deforestation, environmental degradation, and high CO_2 emissions. Charcoal from native forests emits up to nine times more CO_2 than coal. To make 4.7 million tons of charcoal for iron and steel in Brazil, between 9600 and 15,000 square miles (25,000–40,000 km) of forests are cut down every year (Albuquerque 2019; Nogueira and Coelho 2007; Sonter et al. 2015; Kato et al. 2005; Reuters 2007). Not incidentally, making charcoal is one of the most miserable jobs on earth, with a high rate of injuries and often done with slave and child labor.

I would not bet the house on biomass as a fuel for manufacturing. It simply will not scale up. Replacing coal with biomass charcoal would require 36.7 billion tons of wood harvested from 55 million square miles, more land than there is on earth (Smil 2015). In 1810, the US converted 1 million tons of wood from 1500 square miles of forests into charcoal to smelt 49,000 tons of iron. In 2018, the world made 1.8 billion tons of iron with one billion tons of coal.

Pure biomass charcoal can not be used in large modern cement kilns and blast furnaces, though up to 25% biomass can be used to substitute for coke in steel-making. But it is energy-intensive to do so. Trees are cut down, chipped into tiny pieces, trucked to a storage bin, dried out to get rid of the 50% moisture content, pyrolyzed at 850–1475 °F (450–800 °C) which reduces the energy content by 80%, and then are finally transported to the industrial facility (Jahanshahi et al. 2015).

Even after all that processing, biomass charcoal will make poor quality cement and steel, because compared to coke and coal, biomass has a low calorific value, low density, low grindability, low energy density, low combustion efficiency, high moisture content, high volatile content, high bulk volume and is expensive to transport and store. These poor qualities further require pretreatment and mechanical, chemical, or thermal upgrading to improve biomass enough to use in blast furnaces (Pimchuai et al. 2010; Pandey et al. 2015).

Coal coke also makes higher quality iron and steel than charcoal because it can support the descending burden, enhance gas distribution and permeability in the shaft, percolate liquid iron, and more. No other material can achieve the required balance between gases and solids in the blast furnace. Chemically, coke reduces gas for indirect reduction, directly reduces hard oxides, and carburizes hot metal. Thermally coke is unique too. These physical, chemical, and thermal properties can not be totally replaced by biomass charcoal (Mousa et al. 2016).

Biomass can make far less cement and steel than coal. More than a century ago, we realized that cutting down our forests to make charcoal would be ruinous. Circa 1880, biomass charcoal became illegal in the US in order to save forests from being destroyed (Mousa et al. 2016).

One day, cement manufacturing will be limited not just by the availability of high heat, but for a lack of sand. Today, we use more sand than any other resource except air and water, nearly 50 billion tons of sand and gravel a year. Without sand, there would be no concrete, ceramics, computer chips, glass, plastics, abrasives, paint, asphalt pavement, and many other products. Beach sand is required, desert sand is too polished. All beaches may be gone by 2100 due to rising sea levels, increasing numbers of severe storms, massive erosion along coastlines, exploitation for development, and dams trapping sand (Beiser 2018; Gillis 2014). That does not seem possible. Let us hope it is not.

Conclusion

Tick-tock. Promising approaches for using alternative energy for most manufacturing processes and for generating and storing high heat are few and far between. It would take decades and trillions of dollars to scale up alternative energy enough to provide power even if electrical or hydrogen processes existed.

As fossil fuel supplies become harder to come by, manufacturers will simply move to where they can get them, just as back in Wood World when metal smelters and pottery makers used to move to forested areas after depleting their own nearby forests (Perlin 2005).

As fossil energy declines, so too will manufacturing. Oil, coal, and natural gas will be allocated to more important uses such as agriculture, transportation, and electricity generation. There will be less energy for repairs and replacements. Industries will have to be near dwindling supplies of fossil fuels, forests, or ports where energy resources can be imported.

The future will not be as resource rich as in today's world. We will face tough choices. Consuming less is one of the essential ways to cope with a simpler future. We should use this opportunity to mandate objects be reusable, recyclable, and long-lasting as possible. Pollution will be reduced. No more planned obsolescence.

References

Albuquerque J. 2019. Charcoal production: challenges and opportunities. Inflor.com. https://www.inflor.com.br/en/charcoal-production-challenges-and-opportunities/. Accessed 2 Nov 2020

AM (2020) Climate action in Europe. Our carbon emissions reduction roadmap. ArcelorMittal. https://corporate-media.arcelormittal.com/media/yw1gnzfo/climate-action-in-europe.pdf. Accessed 2 Nov 2020

Andrews R (2015) Renewable energy storage and power-to-methane. Energy Matters. http://euan-mearns.com/renewable-energy-storage-and-power-to-methane/. Accessed 2 Nov 2020

Bardi U (2014) Extracted: how the quest for mineral wealth is plundering the planet. Chelsea Green Publishing

Barnard M (2019) Heliogen is Bill Gates' latest venture that is only good for oil & gas Part 1. https://cleantechnica.com/2019/11/22/heliogen-is-bill-gates-latest-venture-that-is-only-good-for-oil-gas-part-1/. Accessed 2 Nov 2020

Beiser V (2018) The world in a grain. The Story of Sand and How It Transformed Civilization. Riverhead Books

Brown JM (2016) How do blast furnaces work. Financial times. https://www.ft.com/content/bc91954a-fb1c-11e5-b3f6-11d5706b613b. Accessed 2 Nov 2020

Brueske S, Sabouni R, Zach C, et al (2012) U.S. manufacturing energy use and greenhouse gas emissions analysis. Energetics incorporated for the Oak Ridge National Laboratory

Cooper M (2013) Renaissance in reverse: Competition pushes aging U.S. nuclear reactors to the brink of economic abandonment. Institute for Energy and the environment. Vermont school of law. http://large.stanford.edu/courses/2018/ph241/shi1/docs/cooper.pdf. Accessed 2 Nov 2020

Dieter CA, Maupin MA, Caldwell RR, et al (2015) Estimated use of water in the United States in 2015. U.S. Geological Survey

DOE (2004). Energy use, loss and opportunities analysis: U.S. manufacturing & mining. U.S. Department of energy

DOE (2014) Manufacturing Energy and carbon footprints: all manufacturing. U.S. Department of Energy. https://www.energy.gov/eere/amo/manufacturing-energy-and-carbon-footprints-2014-mecs

DOE (2015a) Industrial process heating – Technology assessment. U.S. Department of Energy

DOE (2015b) Manufacturing energy and carbon footprint. Sector: All manufacturing. U.S. Department of energy. https://www.energy.gov/eere/amo/manufacturing-energy-and-carbon-footprints-2014-mecs. Accessed 2 Nov 2020

ECRA. 2007. Carbon capture technology – options and potentials for the cement industry. European Cement Research Academy. https://ecra-online.org/fileadmin/redaktion/files/pdf/ECRA_Technical__Report_CCS_Phase_I.pdf. Accessed 2 Nov 2020

EIA (2013) Electricity use by machine drives varies significantly by manufacturing industry. U.S. Energy Information Administration

EIA (2019a) International energy outlook 2019. U.S. Energy Information Administration, Office of Energy Analysis

EIA (2019b) Use of energy explained. Energy use in industry. U.S. Energy Information Administration

EIA (2020a) Net generation by Energy Source Table 1.1, Net Generation from Renewable sources Table 1.1.A. U.S. Energy Information Administration

EIA (2020b) Table 6.07.B. Capacity factors for utility scale generators primarily using non-fossil fuels. U.S. Energy Information Administration

EPA (2017) Industrial process heat technologies and applications. U.S. Environmental protection Agency. https://www.epa.gov/rhc/industrial-process-heat-technologies-and-applications-text-version-diagram. Accessed 2 Nov 2020

Frangoul A (2020) Energy major announces plans to produce hydrogen from 'largest plant of its kind'. CNBC. https://www.cnbc.com/2020/07/02/energy-major-to-produce-hydrogen-from-largest-plant-of-its-kind.html. Accessed 2 Nov 2020

Friedmann SJ, Fan Z, Tang K (2019) Low-carbon heat solutions for heavy industry: sources, options, and costs today. Center on Global Energy Policy at Columbia University SIPA, New York

Gillis JR (2014) Why Sand Is Disappearing. New York Times

Heinberg R, Fridley D (2016) Our renewable future. Island Press

Honsberg C, Bowden S (2019) Refining Silicon. https://www.pveducation.org/pvcdrom/manufacturing-si-cells/refining-silicon. Accessed 2 Nov 2020

Huang E (2019) A hydrogen fueling station explosion in Norway has left fuel-cell cars nowhere to charge. https://qz.com/1641276/a-hydrogen-fueling-station-explodes-in-norways-baerum/. Accessed 2 Nov 2020

Jahanshahi S, Mathieson JG, Somerville MA et al (2015) Development of low-emission Integrated Steelmaking Process. Journal of Sustainable Metallurgy 1:94–114

Kahn JP (2013) How beer gave us civilization. New York Times

Kato M, DeMarini DM, Carvalho AB, et al (2005) World at work: Charcoal producing industries in northeastern Brazil. Occupational & Environmental Medicine. https://doi.org/10.1136/oem.2004.015172

Malico I, Pereira RN, Gonçalves AC et al (2019) Current status and future perspectives for energy production from solid biomass in the European industry. Renew Sust Energ Rev 112:960–977

McDonnell A (2020) Steelmaker ArcelorMittal warns decarbonization of industry would cost billions. The Epoch Times. https://www.theepochtimes.com/steelmaker-warns-decarbonization-of-industry-would-cost-billions_3403309.html. Accessed 2 Nov 2020

Mousa E, Wang C, Riesbeck J et al (2016) Biomass applications in iron and steel industry: an overview of challenges and opportunities. Renew Sust Energ Rev 65:1247–1266

Moya EZ (2012) Parabolic-trough concentrating solar power (CSP) systems. Concentrating solar power technology. Woodhead Publishing, pp 197–239. https://doi.org/10.1533/9780857096173.2.197

Munnion O (2020) Climate finance for charcoal production in Brazil is fueling conflicts with communities. https://www.germanclimatefinance.de/2020/05/25/climate-finance-for-charcoal-production-in-brazil-is-fueling-conflicts-with-communities/. Accessed 5 Dec 2020

Nogueira LAH and Coelho ST (2007) Sustainable charcoal production in Brazil. United Nations Food & Agriculture Organization. http://www.fao.org/3/i1321e/i1321e04.pdf. Accessed 2 Nov 2020

NREL (2013) Concentrating solar power projects. National Renewable Energy Laboratory, Colorado. https://solarpaces.nrel.gov/by-country/US. Accessed 2 Nov 2020

Oberhaus D (2019) A solar 'breakthrough' won't solve cement's carbon problem. Wired. https://www.wired.com/story/a-solar-breakthrough-wont-solve-cements-carbon-problem/. Accessed 2 Nov 2020

Palmer G (2020) Energy storage and civilization: a systems approach. Springer

Pandey A, Negi S, Binod P et al (2015) Handbook of pretreatment of biomass processes and technologies. Elsevier

Perlin J (2005) A forest journey: the story of wood and civilization. Countryman Press

Pimchuai A, Dutta A, Basu P (2010) Torrefaction of agriculture residue to enhance combustible properties. Energy Fuel 24:4638–4645. https://doi.org/10.1021/ef901168f

Pooler M (2019) A greener steel industry still looks a long way off. Replacing coke with hydrogen would remove CO2 emissions – but process remains in infancy. Financial Times. Accessed 2 Nov 2020

Reuters (2007) Slave labor persists in Amazon charcoal works: ICC. https://www.reuters.com/article/us-brazil-slaves/slave-labor-persists-in-amazon-charcoal-works-icc-idUSN2743662520070827. Accessed 2 Nov 2020

Reuters (2019a) Samsung electronics chip output at South Korea plant partly halted due to short blackout. https://www.reuters.com/article/us-samsung-elec-plant/samsung-electronics-chip-output-at-south-korea-plant-partly-halted-due-to-short-blackout-idUSKBN1Z01K3. Accessed 2 Nov 2020

Reuters (2019b) Explosions and subsidies: Why hydrogen is struggling to catch on in Korea. Accidents and infrastructure are holding it back

Sandalow, D, Friedmann J, Aines R, et al (2019) ICEF Industrial heat decarbonization roadmap. Innovation for Cool Earth Forum. https://www.icef-forum.org/roadmap/. Accessed 2 Nov 2020

SBC (2014) Hydrogen-based energy conversion. SBC Energy Institute, California

Smil V (2015) Power density: a key to understanding energy sources and uses. MIT Press

Sonter LJ, Barrett DJ, Moran CJ et al (2015) Carbon emissions due to deforestation for the production of charcoal used in Brazil's steel industry. Nat Clim Chang 5:359–363

Szymkowski S (2019) Following hydrogen facility explosion, fuel-cell vehicle owners left stranded. The explosion happened in June, but some owners have been forced to park their cars due to lack of fuel. cnet.com

WCA (2020) How is steel produced? World Coal Association

West DM, Lansang C (2018) Global manufacturing scorecard: How the US compares to 18 other nations. Brookings. https://www.brookings.edu/research/global-manufacturing-scorecard-how-the-us-compares-to-18-other-nations/. Accessed 2 Nov 2020

Chapter 10
What Alternatives Can Replace Fossil-Fueled Electricity Generation?

We have been talking about renewable energy in the US for decades. Replacing fossil fuels. Where are we right now? How much progress have we made? Let us look at the production of electricity, where renewables have the best opportunity because we are replacing one form of electrical generation with another.

In 1985, 72.5% of the US electric grid was powered by natural gas, coal, and oil (EIA 2019). In 2019, it was still nearly two-thirds of fossil fuels. As of 2020, wind and solar contribute just 9%of our electricity (EIA 2020). If wind and solar are to grow up to become energy titans, the time is ripe for their growth spurts. That will require a lot of scaling up, especially since they will also need to fill in for nuclear power, which is likely to decline from 19.7% to 10% of generation as older plants retire. Natural gas is playing the largest role in replacing coal plants so far, not wind and solar (Paraskova 2020).

Ironically, the one time we "appreciate" our electrical grid is when the lights go out. The US electric grid is the largest machine in the world. With 200,000 miles of high-voltage transmission lines and 5.5 million miles of local distribution lines, it streams electricity from thousands of power plants to millions of homes, factories, and businesses. The grid is like a circus tightrope walker. Supply and demand must exactly balance or the tightrope walker falls off and the grid blacks out. Too little power and the lights flicker and industrial machines seize. Too much and substation transformers melt.

In the morning, as people awake, natural gas power plants ramp up, and instantly step in and out as wind and solar surge and die, maintaining balance. They also supply and supplement power on a few days of peak demand. Natural gas and coal serve as the backup storage for wind and solar, since there is not nearly enough battery, pumped hydro, or other storage to keep the grid up (GTM 2019). Like natural gas, hydropower can dispatch electricity quickly as well, though not year-round.

A. J. Friedemann, *Life after Fossil Fuels*, Lecture Notes in Energy 81, https://doi.org/10.1007/978-3-030-70335-6_10

Power Players: Job Applicants to Replace Fossil-Fuel Generated Electricity

Offshore wind (offshore of land) is commercial. So far, they are mainly in Europe. The US has one small wind farm of five turbines off Rhode Island that cost $290 million to build to supply the 1000 residents of Block Island with electricity. That is 290 thousand dollars per customer. Who paid for that? Two new turbines were recently constructed near Virginia that can power 3000 homes at peak output.

In the US, 90% of the windiest offshore sites are in water too deep for current technology. Offshore turbines are too large to travel on land and must be launched from new $1.3 billion port facilities with $100–250 million vessels capable of lifting 757-ton turbines (Navigant 2013; NREL 2010).

Offshore wind turbines operate in a brutal environment and have an average lifetime of just 15 years because of rust and damage from tides, storms, hurricanes, lightning, icing and icebergs, large waves, marine growth, and corrosion (Navigant 2013). Given their high cost and short lifespan, no wonder so few have been built.

Nuclear Power. As of 2020, only two new nuclear reactors are being built in the US. Why? Probably not so much because of safety and environmental concerns, but economics. They are not economically competitive, costing $8.5–$20 billion and requiring 10 years to construct. Not many banks are willing to loan billions of dollars to an enterprise that will not generate a dollar of income for 10 years.

A natural gas plant equal in power to a nuclear plant can be built for $2.5 billion in half the time. This is why in the US there are only 96 nuclear reactors at 58 sites left in the US. Nine of them are scheduled to shut down because they are old, uneconomic, unreliable, subject to breakdowns, have long outages, safety issues, and require mandatory expensive post-Fukushima retrofits (Cooper 2013). Since 1963, 36 reactors have shut down in the US. If their costs and construction times were equivalent, nukes still would be at a disadvantage: Opposition to new nuclear plants is strong because, among other issues, there is no permanent waste storage site (Alley and Alley 2013). There are also concerns about the proliferation of nuclear bomb-grade materials. Gen IV, breeder, and thorium reactors, in principle cheaper and safer, remain in the research stage and are far from being commercial.

As for nuclear power plants serving the role of quickly stepping in when solar and wind are not generating, nukes can not perform the tightrope act because they require up to 8 h to ramp up or down. Thus they, like coal plants, are considered "base load."

Additionally, uranium is not renewable. Someday there will be a peak uranium (Dittmar 2013; EWG 2013; Bardi 2014). Extraction from seawater is not an option. That requires too much energy (Bardi 2014).

The average age of US nuclear reactors is 38 years old. Accidents are increasingly likely after 20 years (Hirsch et al. 2007). These plants were initially licensed for just 40 years, though extensions are usually granted.

The Nuclear Reactor commission asked the operators of 60 US nuclear power plant sites to model their current flood risk, including the likely effects of climate change on their mostly 50+ year old plants. If they were not built to withstand potential floods, then plans to remedy future flooding needed to be provided. Ninety percent of these reactor sites need to be modified, with 54 having at least one flood risk exceeding their design, 53 not built to withstand the current risk from intense precipitation, 25 in jeopardy based on current flood projections for streams and rivers, and 19 that were not designed for their maximum storm surge (Flavelle and Lin 2019).

As is well understood, nuclear power plants are unique in the scale of risk they pose to society. In Pennsylvania, if the Peach Bottom nuclear spent fuel pool caught on fire, up to 8.8 million people might have to evacuate, aquifers would be contaminated, and thousands of square miles of land would become uninhabitable (von Hippel and Schoepnner 2016; Michaelides 2012; Lyman et al. 2017).

Fusion. Nuclear physicists often say—it is a rueful refrain—that "fusion is 30 years away … and always will be." Despite research since the 1940s, fusion energy remains a very distant hope (Cartlidge 2017; Moyer 2010; NRC 2013a; Perlman 2014). Fusion is the only energy resource with the theoretical potential to scale up enough to replace fossil fuels (Hoffert et al. 2002).

Hydropower is quickly dispatchable like natural gas and the best way to restart a grid when it crashes. However, hydro is concentrated in just 10 states that have 80% of hydropower, with Washington state a whopping 25% of that (HS 2017). These 10 states can not use hydropower year-round due to drought, water storage for cities and agriculture, fisheries, and ecosystem protection. Nor can more dams be built; there are few places to put them. Dams are not renewable since their average lifespans are 60–80 years. By 2025, 70% of the dams in the US will be over 50 years old (ASCE 2017). Dams can fail in extreme floods when their reservoirs fill with sediment, or their concrete erodes. Climate change will further reduce hydropower. It already has—the southwest is in the worst mega-drought in 500 years and one of the most severe in history (MacFarlane 2018). The Hoover Dam has been downrated from 2074 Megawatts to 1592 as water levels continue to drop in Lake Mead.

Geothermal provides just 0.4% of US power, with two states producing 96% of it: California 81% and Nevada 15%. Access to geothermal is possible only in a few rare, near-surface geologic areas with a lot of water, power lines, and roads nearby. Untapped sites are usually too small to justify transmission lines, too expensive, or lack water (CCST 2011; Hurlbut 2012; NREL 2013; Rhyne et al. 2015).

Concentrated Solar Power (solar thermal) plants, which use an array of mirrors or lenses to focus and concentrate beams of sunlight, cost about a billion dollars each and generate just 0.06% of US electricity. Just as two-thirds of the energy in coal and natural gas are lost as heat when converted to electricity, the same is true at a CSP plant, where two-thirds of energy is lost when water is turned to steam and then converted from steam to electricity (De Decker 2011).

Marine Hydro-kinetic (MHK) power is far from commercial because variable wave, tidal, and ocean current energy capturing machines rust and are destroyed by the same problems offshore wind turbines face. They are also expensive to build and maintain and have low energy efficiency. It is hard to find a place to put them since they have to be near urban centers and the electric grid, yet not conflict with ship navigation, aquaculture, marine sanctuaries, and nearby ports. Hydro-kinetic devices are huge. To generate 1000 MW of wave power in the rough North Sea, the Wave Dragon Energy Converter would need to be 124 miles long (NRC 2013b). The prototype was scrapped in 2011.

That Leaves Photovoltaic Solar and Onshore Wind to Save the Day

The vast majority of a 100% decarbonized grid needs to be wind and solar, backed up with flexible natural gas (GTM 2019).

But vast geographic areas do not have wind or solar. Like hydropower, just 10 states produce 75% of wind power in the US (EIA 2017), and just 10 states produce 79% of solar power (CE 2020). Wind is seasonal, with very little wind potential across the entire US in the summer, or any time of year in the Southeast (NREL 1983). Solar power is seasonal too, weak over much of the US half the year and strong—sometimes too strong!—for 3 months mainly in the southwest (NREL 1991).

Fossil fuel and nuclear plants are available nearly all the time. Solar and wind are intermittent. Much of the time, even in the states fortunate to have good wind and solar, you can not count on the sun and wind to provide power when needed. In the Midwest, which has some of the strongest winds in the US, MISO is the organization that manages a 15-state electric grid. MISO (2018) reports it is able to firmly rely on and power the entire grid with wind 15.6% of the time. The remaining 84.4% of the time, it must supplement what the wind provides with natural gas, coal, or nuclear backup.

There is no national grid for sharing electricity (St. John 2020), but even if one existed, the few times of the year when the lucky states produced more electricity than their own manufacturing, business, and residential customers could use, there would be little left to share.

Many of the fortunate states have already built out wind and solar on their prime sites (NREL 2013). Adding more in secondary or remote locations is rarely cost-justified and can require billions of dollars of new transmission lines, development often fiercely fought by landowners.

Wind Turbines and Solar Panels Are REBUILDABLE, Not RENEWABLE

Mining consumes 10% of world energy (TWC 2020). Wind, solar, and all other electrical generating machines rely on fossil-fueled mining, manufacturing, and transportation every step of their life cycle.

Consider a wind turbine. Let us build a windmill: Mining trucks and machinery dig into the earth for iron ore, ships or rail carry ore to a coke-fired smelter or blast furnace to extract the metal. The 8000 parts of a wind turbine, all made with hydrocarbon energy and materials, are delivered from all over the world to the assembly factory. Coal creates cement, and is transported by cement trucks to pour tons of concrete and steel rebar for the turbine platform. Several turbine segments are driven to the installation site, workers arrive in gassed up vehicles, the whole parade of them over roads paved with asphalt, and a crane lifts segments of wind turbines into place (Roberts 2019).

Clearly, after their 20-year lifespan, wind turbines can not be replaced (Hughes 2012; Davidsson et al. 2012, 2014) or solar panels every 18–25-year lifespan (Ferroni and Hopkirk 2016; Shaibani 2020) without fossil-fueled or electrified transportation and manufacturing.

There Is Not Time, Energy, or Materials to Make So Many Rebuildable Contraptions

Solar and wind require more than clear skies and stiff breezes. They require rare earth and other scarce minerals. These same minerals are also used in electric vehicles, electronics, computers, the power grid, phones, magnets, electric motors, satellites, semiconductors, batteries, lasers, telecommunications, fuel cells, fiber optics, catalysts, integrated circuits, GPS navigation, and as alloys in steel and aluminum.

Computers are an essential tool used to manage the electric grid and monitor wind and solar power. They are made of 60 minerals, many quite rare with no substitutable element (NRC 2008; Graedel et al. 2015; EC 2017). Fortunately, an abacus can be made entirely with renewable wood.

By mid-century, minerals and metals needed for high-tech could be running short, including stainless steel, copper, gallium, germanium, indium, antimony, tin, lead, gold, zinc, strontium, silver, nickel, tungsten, bismuth, boron, fluorite, manganese, selenium, and more (Kerr 2012, 2014; Barnhart and Benson 2013; Bardi 2014; Veronese 2015; Sverdrup and Olafsdottir 2019; Pitron 2020 Appendix 14).

Based on projected growth of solar and wind, by 2050 wind turbines and solar panels will need 12 times as much indium as the entire world produces now, seven times more neodymium, and three times more silver (Van Exter et al. 2018). Solar panels that use tellurium, germanium, and ruthenium also face shortages (Grandell and Höök 2015).

Bardi (2014) wrote, "The limits to mineral extraction are not limits of quantity; they are limits of energy. Extracting minerals takes energy, and the more dispersed the minerals are, the more energy is needed. Today, humankind doesn't produce sufficient amounts of energy to mine sources other than conventional ores, and probably never will."

Recycling appears to be a solution. But some elements are impossible to recover from composite materials and alloys (Bloodworth 2014; Izatt and Hageluken 2016). The cost to recover most rare metals exceeds their value. Recycling is time-consuming and uses toxic chemicals to separate minerals apart. Of the 60 most used industrial metals, 34 are recycled less than 1% of the time, and another eight less than half the time (UNEP 2011).

Sunshine may be eternal, but solar panels do not last forever. Ninety percent of solar panels that have reached the end of their lives are going into the landfill in the US This is much cheaper than recycling, which requires dissembling, etching, and melting panels to remove lead, cadmium, copper, gallium, indium, aluminum, glass, and silicon solar cells.

And so too with wind turbines. They will not spin free forever and they are not fully recyclable, so 720,000 tons of blade material are expected to end up in landfills over the next 20 years. Especially the up to 300-foot-long blades made of resin and fiberglass, materials not worth salvaging (Stella 2019).

For some minerals, a single country produces more than half, and three nations control 75% of lithium, cobalt, and some rare earths (Veronese 2015). China controls about 63% of rare earth rare metals, but more importantly, about 90% of the supply chain, from extraction, processing separation, and refining, to manufactured goods. If new mines were built to lessen dependence, it would take about 15 years to construct one, and most of the production would probably be sold to China for processing and products made there (GAO 2010; AI 2019; Hui 2020).

But why compete? Let China monopolize the second most polluting industry on earth. Mining spews out acid rain, wastewater, and heavy metals onto land, water, and air (PEBI 2016). One-fifth of China's arable land is polluted from mining and industry (BBC 2014). Mining the materials needed for renewable energy potentially affects 50 million square kilometers, 37% of Earth's land (minus Antarctica), with a third of this land overlapping key biodiversity areas, wilderness, or protected areas. If mined, that would drive biodiversity loss, harm (rain) forests, and poison ecosystems (Kleijn et al. 2011; Hickel 2019; Sonter et al. 2020).

Renewable energy is anything but clean and green. And quite a Pyrrhic victory for China!

Biomass Electric Power

Biomass. This is the only commercial, renewable energy that can generate electricity around the clock and balance intermittent wind and solar power (though not within seconds to minutes). This is especially applicable in the 10 states that have the most forest cover to fuel biomass power plants.

Chapter 29 will discuss using biomass to generate electricity. Chapter 29? Okay, I know that is a long wait, I know that delayed gratification is not your bag and that the anticipation is killing you. But trust me, I have a plan. You are not ready for it yet. Follow me to Chapter 11.

References

AI (2019) Rare earth elements: market issues and outlook, Adamas Intelligence. https://www. adamasintel.com/rare-earth-market-issues-and-outlook/. Accessed 2 Nov 2020

Alley WM, Alley R (2013) Too hot to touch: the problem of high-level nuclear waste. Cambridge University Press

ASCE (2017) America's infrastructure report card. Dams D. American Society of Civil Engineers. https://www.infrastructurereportcard.org/cat-item/dams/. Accessed 2 Nov 2020

Bardi U (2014) Extracted: how the quest for mineral wealth is plundering the planet. Chelsea Green

Barnhart CJ, Benson SM (2013) On the importance of reducing the energetic and material demands of electrical energy storage. Energy Environment Science 2013(6):1083–1092

BBC (2014) Report: One fifth of China's soil contaminated. https://www.bbc.com/news/world-asia-china-27076645. Accessed 2 Nov 2020

Bloodworth A (2014) Track flows to manage technology-metal supply. Recycling cannot meet the demand for rare metals used in digital and green technologies. Nature 505:19–20

Cartlidge E (2017) Fusion energy pushed back beyond 2050. BBC. https://www.bbc.com/news/science-environment-40558758. Accessed 2 Nov 2020

CCST (2011) California's energy future: the view to 2050 summary report. California Council on Science & Technology

CE (2020) Solar energy generation by state. ChooseEnergy.com. https://www.chooseenergy.com/data-center/solar-energy-production-by-state/. Accessed 2 Nov 2020

Cooper M (2013) Renaissance in reverse: competition pushes aging U.S. nuclear reactors to the brink of economic abandonment. Institute for Energy & the environment, Vermont Law School

Davidsson S, Höök M, Wall G (2012) A review of life cycle assessments on wind energy systems. Int J Life Cycle Assess 17:729–742

Davidsson S, Grandell L, Wachtmeister H et al (2014) Growth curves and sustained commissioning modelling of renewable energy: Investigating resource constraints for wind energy. Energy Policy 73:767–776

De Decker K (2011) The bright future of solar thermal powered factories. Low Tech magazine. https://www.lowtechmagazine.com/2011/07/solar-powered-factories.html. Accessed 2 Nov 2020

Dittmar M (2013) The end of cheap uranium. Sci Total Environ 461–462:792–798

EC (2017) Communication from the commission to the European Parliament on the 2017 list of critical raw materials for the EU. European Commission, Brussels

EIA (2017) Wind turbines provide 8% of U.S. generating capacity, more than any other renewable source. U.S. Energy Information Administration

EIA (2019) Table 7.2a Electricity net generation total (all sectors). U.S. Energy Information Administration

EIA (2020) Net generation by Energy Source Table 1.1, Net Generation from Renewable sources Table 1.1.A. U.S. Energy Information Administration

EWG (2013) Fossil and nuclear fuels – the Supply Outlook. Energy Watch Group

Ferroni F, Hopkirk RJ (2016) Energy Return on Energy Invested (ERoEI) for photovoltaic solar systems in regions of moderate insolation. Energy Policy 94:336–344

Flavelle, C, Lin JCF (2019) U.S. Nuclear power plants weren't built for climate change. Bloomberg

GAO (2010) Rare earth materials in the defense supply chain. U.S. Government Accountability Office

Graedel TE, Harper EM, Nassar NT et al (2015) On the materials basis of modern society. US Proceedings of the National Academy of Sciences 112:6295–6300

Grandell L, Höök M (2015) Assessing rare metal availability: challenges for solar energy technologies. Sustainability 7:11818–11837

GTM (2019) Why Flexible Gas Generation Must Be Part of Deep Decarbonization. Greentechmedia. com. https://www.greentechmedia.com/articles/read/why-flexible-gas-must-be-part-of-the-path-to-100-percent-decarbonization. Accessed 2 Nov 2020

Hickel J (2019) The limits of clean energy. If the world isn't careful, renewable energy could become as destructive as fossil fuels. Foreign policy. https://foreignpolicy.com/2019/09/06/the-path-to-clean-energy-will-be-very-dirty-climate-change-renewables/. Accessed 2 Nov 2020

Hirsch H, Becker O, Schneider M, et al (2007) Nuclear reactor hazards ongoing dangers of operating nuclear technology in the 21st century. Estudos Avancados. https://doi.org/10.1590/S0103-40142007000100020

Hoffert MI, Caldeira K, Benford G et al (2002) Advanced technology paths to global climate stability: energy for a greenhouse planet. Science 298:981–987

HS (2017) Dams and energy sectors interdependency study. An update to the 2011 study. U.S. Department of Energy and Homeland Security

Hughes G (2012) The performance of wind farms in the United Kingdom and Denmark. Renewable Energy Foundation. https://www.ref.org.uk/attachments/article/280/ref.hughes.19.12.12.pdf. Accessed 3 Nov 2020

Hui M (2020) How China built up its dominance in rare earths. qz.com. https://qz.com/1924282/how-china-became-dominant-in-rare-earths/. Accessed 2 Nov 2020

Hurlbut D (2012) Geothermal power and interconnection: The economics of getting to market. National Renewable Energy Laboratory, Golden

Izatt RM, Hageluken C (2016) Chapter 1 Recycling and sustainable utilization of precious and specialty metals pp 1–22, in Metal Sustainability: Global Challenges, Consequences, and Prospects. John Wiley & Sons, Ltd.

Kerr RA (2012) Is the world tottering on the precipice of peak gold? Science 335:1038–1039

Kerr RA (2014) The coming copper peak. Science 343:722–724

Kleijn R, Van der Voet E, Kramer GJ et al (2011) Metal requirements of low-carbon power generation. Energy 36:5640–5648

Lyman E, Schoeppner M, von Hippel F (2017) Nuclear safety regulation in the post-Fukushima era. Science 356:808–809

MacFarlane D (2018) The Southwest might be in one of the worst mega-droughts in history. https://weather.com/science/environment/news/2018-12-19-southwest-mega-drought. Accessed 3 Nov 2020

Michaelides EE (2012) Alternative energy sources. Springer

MISO (2018) Planning year 2019–2020. Wind & solar capacity credit. Midcontinent Independent System Operator. https://cdn.misoenergy.org/2019%20Wind%20and%20Solar%20Capacity%20Credit%20Report303063.pdf

Moyer M (2010) Fusion's false dawn. Scientific American

Navigant (2013) U.S. Offshore Wind Manufacturing and Supply Chain Development. U.S. Department of Energy

NRC (2008) Minerals, critical minerals, and the U.S. economy Chapter 2. National Research Council, National Academies Press

NRC (2013a) An Assessment of the Prospects for Inertial Fusion Energy. National Research Council, National Academies Press

NRC (2013b) An Evaluation of the U.S. Department of Energy's marine and hydrokinetic resource assessments. National Research Council, National Academies Press

NREL (1983) Wind Energy Resource Atlas of the United States. National Renewable Energy Laboratory

NREL (1991) U.S. Solar Radiation Resource Maps: Atlas for the solar radiation data manual for flat plate and concentrating collectors. National Renewable Energy Laboratory

NREL (2010) Assessment of offshore wind energy resources for the United States. National Renewable Energy Laboratory, U.S. Department of Energy

NREL (2013) Beyond renewable portfolio standards: An assessment of regional supply and demand conditions affecting the future of renewable energy in the west. Golden: National Renewable Energy Laboratory

Paraskova T (2020) Natural gas has replaced over 100 U.S. coal plants in the last decade. Oilprice.com. https://oilprice.com/Energy/Coal/Natural-Gas-Has-Replaced-Over-100-US-Coal-Plants-In-The-Last-Decade.html

PEBI (2016) World's worst pollution problems. The toxins beneath our feet. Pure Earth Blacksmith Institute. https://www.worstpolluted.org/2016-report.html. Accessed 3 Nov 2020

Perlman D (2014) Livermore Lab's fusion energy tests get closer to 'ignition'. San Francisco Chronicle. https://www.sfgate.com/science/article/Livermore-Lab-s-fusion-energy-tests-get-closer-to-5229592.php. Accessed 3 Nov 2020

Pitron G (2020) The Rare Metals War: The Dark Side of Clean Energy and Digital Technologies. Scribe US

Rhyne I, Klein J, Neff B (2015) Estimated cost of new renewable and fossil generation in California. California Energy Commission

Roberts D (2019) These huge new wind turbines are a marvel. They're also the future. https://www.vox.com/energy-and-environment/2018/3/8/17084158/wind-turbine-power-energy-blades. Accessed 3 Nov 2020

Shaibani M (2020) Solar panel recycling: turning ticking time bombs into opportunities. PV magazine. https://www.pv-magazine.com/2020/05/27/solar-panel-recycling-turning-ticking-time-bombs-into-opportunities/. Accessed 3 Nov 2020

Sonter LJ, Dade MC, Watson JEM et al (2020) Renewable energy production will exacerbate mining threats to biodiversity. Nat Commun 11:4174

St. John J (2020) Transmission emerging as major stumbling block for state renewable targets. Greentechmedia. https://www.greentechmedia.com/articles/read/transmission-emerging-as-major-stumbling-block-for-state-renewable-targets. Accessed 3 Nov 2020

Stella C (2019) Unfurling the waste problem caused by wind energy. National Public Radio. https://www.npr.org/2019/09/10/759376113/unfurling-the-waste-problem-caused-by-wind-energy. Accessed 3 Nov 2020

Sverdrup HU, Olafsdottir AH (2019) Assessing the long-term global sustainability of the production and supply for stainless steel. Biophysical Economics and Resource Quality 4:8

TWC (2020) Energy use from mining. The World Counts. https://www.theworldcounts.com/challenges/planet-earth/mining/energy-use-in-the-mining-industry/story. Accessed 3 Nov 2020

UNEP (2011) Recycling rates of metals. A status report. United Nations Environmental Programme

Van Exter P, Bosch S, Schipper B, et al (2018) Metal demand for renewable electricity generation in the Netherlands. Universiteit Leiden

Veronese K (2015) The high-stakes race to satisfy our need for the scarcest metals on earth. Prometheus Books

von Hippel FN, Schoepnner M (2016) Reducing the danger from fires in spent fuel pools. Science & Global security 24:141–173

Chapter 11
Energy Storage: Excess Electricity from Solar and Wind Must Be Stored

Fossil fuels are energy storage. Say what? Okay, I owe you an explanation.

There is very little electricity stored now because with fossil fuels there has been no need for it. Which is a good thing as discerning readers will soon understand. The coal and natural gas that generate two-thirds of electricity, and nuclear uranium that generates 20% of power *are* the energy storage, and have provided many decades of energy storage so far. Each fossil plant has storage too. If weather or a derailed train prevent a coal delivery, no problem, up to 3 months of coal are stored at coal plants, and 2 months of natural gas is in underground pipelines.

In comparison, wind and solar electricity are intermittent. Sometimes there is too little and sometimes there is more than the grid can use. The extent to which intermittent wind and solar can supplant fossil fuels will depend on storage. Without storage, the grid blacks out. Today excess generation is curtailed, but it will be essential for renewable grids to store excess power to kick in when the wind and sun are not there and to prevent transformers from melting, catching on fire, and even exploding. How much excess energy can wind and solar generate and how much of that can be stored?

A National Grid

Although it seems like the sun must be shining and the wind blowing somewhere, that is not sufficient. Even a national grid sending electric sunshine from California to frosty Illinois and electric Midwest wind to Alabama would need energy storage. Summer sun and winter wind are seasonal.

So, a national grid does not get rid of the need for storage. It is unlikely our regional grids will be combined into a national grid because of the potential for a nationwide blackout from cyber-attacks, terrorism, or aging equipment. Additionally, a national grid might be less stable (Blumsack 2006; Brown et al. 2008).

© The Author(s), under exclusive license to Springer Nature Switzerland AG 2021
A. J. Friedemann, *Life after Fossil Fuels*, Lecture Notes in Energy 81,
https://doi.org/10.1007/978-3-030-70335-6_11

The existing regional grids are badly in need of maintenance and upgrade (Makansi 2007; Munson 2008; ASCE 2017). PG&E in California needs up to $150 billion and 650,000 workers just to clear trees and vegetation near power lines to prevent wildfires (Serna and Rainey 2019). So where would the tens of trillions of dollars to nationalize the grid come from? Regionally, NIMBYism and bureaucratic challenges have caused decade-long delays in expanding the grid, or stopped projects entirely (DOE 2002; Smil 2010).

Seasonal Energy Storage

The sheer magnitude of energy storage required to bring about a 100% renewable grid is a daunting challenge. A study in Europe found that even with a giant supergrid across the European Union, North Africa, and the Mediterranean, a much larger and sunnier area than the United States, there would still need to be 1 month of energy storage to keep the grid up during seasonal variations (Droste-Franke 2015). Palmer (2020) thought that up to 7 weeks of storage would be required as well as large amounts of renewable overbuild. Seasonality is also a problem in California requiring a great deal of storage (Greenblatt et al. 2012). And that is just the electric grid, 18.9% of our total energy consumption (IEA 2020).

As of now, natural gas power plants that can ramp up and down quickly are essential for balancing the intermittency of wind and solar because it takes hours for coal and nuclear plants to ramp up or down. What are the other possibilities? I now present you with the leading candidates. Promising or pipedream—you be the judge.

Pumped Hydro Storage

Pumped hydro storage (PHS) and hydroelectric can back up natural gas and quickly balance the grid when water is available.

PHS is commercial right now and generates power using electric turbines to move water uphill at night when demand is low to a reservoir above. When demand is high, the water is released to flow downhill to generate hydroelectricity. Locations must have both high elevation and space for a reservoir above an existing body of water, which is why there are only 43 of them in the US, with just two built since 1995. New locations are limited and can cost hundreds of millions of dollars.

In terms of energy storage, PHS provides 98% of all the electrical energy stored in the world (Mongird et al. 2019). I can hear you getting excited. Sobriety check: In the US PHS generated *negative* 5.26 Terrawatt hours (TWh) because it took 20–30% more energy to pump water uphill than was generated downhill. Gulp!

Pumped hydro may account for almost all electrical storage, but it is just a drop in the bucket. PHS stores only 0.12% of the 4118 TWh annual electricity generated in 2019.

Amazingly, the US has 43 existing pumped hydro plants. Over a year, they have produced as much as 2 days of energy storage (23 TWh). Keep in mind that was done over 365 days. To store just one day of US electricity generation would require 7848 new PHS dams (365/2 * 43). And a month of storage would require 219,730 additional PHS dams.

How could so many pumped hydro dams be needed? PHS has astonishingly little energy density compared to oil. To store the energy contained in just one gallon of gasoline takes 55,000 gallons of water pumped 726 ft high (Greenblatt et al. 2012). To get your head around this, let me put it this way: It would take 16 Mississippi rivers flowing from a reservoir 650 ft above a second reservoir below to provide just half of US peak load (Palmer 2020). Imagine that.

Energy Storage with Electrochemical Batteries

Batteries are commercial and can kick in suddenly, like natural gas, to balance wind and solar. The most promising battery for utility-scale energy storage is sodium sulfur (NaS). Among electrochemical prospects, this is the only kind of battery with enough materials on earth (Barnhart and Benson 2013) to scale up to this job. Most utility-scale storage and electric car batteries are lithium, a shame given that lithium is relatively rare (Vazquez et al. 2010; Vikström et al. 2013; Penn 2018).

As with PHS dams, staggering numbers of batteries would be required in order to provide backup energy to the grid. To back up just one day of US electricity generation, 11.28 TWh, sodium sulfur (NaS) utility-scale batteries would cost $40.77 trillion dollars, cover 923 square miles, and weigh 450 million tons. Lead-acid (advanced) would cost $8.3 trillion dollars, take up 217.5 square miles, and weigh 15.8 million tons. Lithium-ion batteries would cost $11.9 trillion dollars, take up 345 square miles, and weigh 74 million tons (Akhil et al. 2013).

Alternatively, if there were 133 million Tesla Model S cars (roughly one per family) and a smart grid invented, their 85 kWh battery packs could be used. Each pack weighs 1200 pounds (540 kg) with 7104 lithium-ion battery cells inside, for a total weight of 79.8 million tons. Together, these Tesla batteries could store one day of US electricity generation.

In a 100% renewable electric grid, at least a month of storage is needed to cope with low seasonal solar and wind generation, so multiply all the above by 28 days and then some to account for round-trip efficiency. If a lead-acid battery is used, multiply by 36, since only 72% of the energy put in is returned and by 32 for Li-ion or NaS batteries with 85% round-trip efficiency (Mongird et al. 2019).

Then, when the batteries expire after 10–15 years (8 years if using lead-acid batteries), start over again. And despite their numbers and tonnage, batteries are not powerful enough to restart the grid when it comes down (Palmer 2020).

No wonder we have so little electric grid storage. At the end of 2018, the US had 0.0012 TWh of grid battery energy storage, or 9 s of electricity generation (EIA 2019). Got candles?

Underground Compressed Air Energy Storage (CAES) Gas Turbines

Is not it amazing how ingenious we are: Air pumped into and compressed in natural underground chambers, then released when needed to power a turbine. What makes this go: Location, location, location. Alas, not many locations. There is only one commercial CAES facility in the US and one in Germany because there are so few places to put them. They require hollow salt formations in caverns at least 1650–4250 ft deep to compress air into (Hovorka and Nava 2009). Finite natural gas is required to force the air to compress and decompress with a very low round-trip energy efficiency of just 40–52% (SBC 2013). You saw this one coming: Basically, this approach is full of hot air.

Concentrated Solar Power (CSP) with Thermal Energy Storage

Concentrated solar plants are also very limited by geography. They are feasible only in extremely dry regions with no humidity, haze, or pollutants, which is why seven of the eight in the US are in the desert southwest, generating 96% of all CSP power. These plants cost about $1 billion each and sprawl across huge areas. California's Ivanpah $2.3 billion dollar facility takes up 5.5 square miles. They are expensive, but they only generate 0.06% of US electricity, and only a quarter of them have thermal storage to continue to provide power after the sun goes down, their only advantage over solar photovoltaic. The Crescent Dune CSP molten salt energy storage performed so poorly it went bankrupt, probably before it had paid back the fossil fuel used in its construction. Nor are CSPs fossil-free. Ivanpah burns 525 million cubic feet of natural gas a year to warm up, enough to generate 124 GWh of power at a natural gas plant (Wikipedia 2020). Although they are few and far between, CSPs are nice shiny eye candy while driving through parched desert landscapes.

Thermal Energy Storage

Heat can be stored in water, molten salt or metal, and solid rock. But since there is no material that is a perfect insulator, heat is continually lost over a short time, and all heat is lost eventually. High-quality insulation may store heat for up to 12 h, but it does not work for seasonal cycles (Michaelides 2012).

Biomass Energy Storage

Many scientists think the only way to cope with seasonal variations in solar and wind would be hydrogen or power2gas, both far from commercial with many obstacles to overcome.

So once again, biomass is the only commercial renewable energy that could keep the electric grid up. Picture the stockpiles of wood. No smoking allowed around the woodpile!

References

Akhil AA, Huff G, Currier AB, et al (2013) Electricity storage handbook in collaboration with NRECA. Sandia National Laboratories and Electric Power Research Institute

ASCE (2017) America's infrastructure report card. Energy D+. American society of civil engineers

Barnhart CJ, Benson SM (2013) On the importance of reducing the energetic and material demands of electrical energy storage. Energy & Environment Science 6:1083–1092

Blumsack SA (2006) Network topologies and transmission investment under electric-industry restructuring. Carnegie Mellon University, Pittsburgh

Brown M, Cibulka L, Miller L, et al (2008) Transmission technology research for renewable integration. California Energy Commission

DOE (2002) National transmission grid study. United States Department of Energy

Droste-Franke B (2015) Review of the need for storage capacity depending on the share of renewable energies (Chap. 6). In Electrochemical energy storage for renewable sources and grid balancing. Elsevier

EIA (2019) Most utility-scale batteries in the United States are made of lithium-ion. U.S. Energy Information Administration. https://www.eia.gov/todayinenergy/detail.php?id=41813. Accessed 4 Nov 2020

Greenblatt J, Long J, Hannegan B (2012) California's energy future: electricity from renewable energy and fossil fuels with carbon capture and sequestration. California Council on Science and Technology, California

Hovorka S, Nava R (2009) Characterization of bedded salt for storage caverns: case study from the Midland Basin. U.S. Department of Energy

IEA (2020) Key world energy statistics. International Energy Agency

Makansi J (2007) Lights out: the electricity crisis, the global economy, and what it means to you. Wiley

Michaelides EE (2012) Alternative energy sources. Springer

Mongird K, Viswanathan V, Balducci P et al (2019) Energy storage technology and cost characterization report. U.S. Department of Energy

Munson R (2008) From Edison to Enron: The business of power and what it means for the future of electricity. Praeger

Palmer G (2020) Energy storage and civilization: a systems approach. Springer

Penn I (2018) How zinc batteries could change energy storage. New York Times

SBC (2013) Electricity storage Factbook. SBC Energy Institute. https://www.cpuc.ca.gov/WorkArea/DownloadAsset.aspx?id=3170. Accessed 4 Nov 2020

Serna J, Rainey J (2019) California's huge, humiliating power outages expose the vulnerabilities of PG&E's power grid. https://www.latimes.com/california/story/2019-10-10/pg-e-california-power-outages-grid-climate-change. Accessed 4 Nov 2020

Smil V (2010) Energy myths and realities. AIE Press

Vazquez S, Lukic SM, Galvan E et al (2010) Energy storage systems for transport and grid applications. IEEE Trans Ind Electron 57:3881–3895

Vikström H, Davidsson S, Höök M (2013) Lithium availability and future production outlooks. Appl Energy 110:252–266

Wikipedia (2020) Ivanpah Solar Power Facility. https://en.wikipedia.org/wiki/Ivanpah_Solar_Power_Facility. Accessed 22 Nov 2020

Chapter 12
Half a Million Products Are Made Out of Fossil Fuels

Ever been out in the suburbs or rural area and seen one of those farm-sized Amazon warehouses? Ever got lost inside a Walmart store? Ever been trapped inside the maze of aisles in an Ikea store? We live in a consumer society. We consume about ONE BILLION TONS of products a year. We live like kings.

Of all the fossil fuels we use in a year, about 7%—500 million metric tons of oil equivalent, the weight of all the people on earth—is used as both feedstock and energy to make these one billion tons of products (IEA 2018). Mostly it is oil for high-value chemicals. Natural gas and coal are used to make ammonia and methanol, but difficult to turn into other products because they require multiple energy-intensive steps (IEA 2018, KAUST 2020).

How do I love thee crude oil? Let me count the ways: Plastic, asphalt, glue, aspirin, insecticides, antiseptics, bandages, purses, boats, cameras, shampoo, candles, cell phones, curtains, luggage, dashboards, fertilizers, ink, pharmaceuticals, refrigerants, shower curtains, surfboards, synthetic rubber, tents, toothpaste, Legos, and my umbrella. And unfortunately, single use plastic bottles.

Seeking 500 Million Tons a Year of Something that can Replace Fossil Fuel Products

Fossil fuels are mostly carbon (C) and hydrogen (H). Natural gas is 20% C and 80% H, petroleum 84% C and 12% H, and coal 84% C and 5% H.

What else on earth is both abundant and chock-a-block with carbon and hydrogen that can replace fossil fuels? Let us start with abundant. The world is mostly made of soil, air, water, and plants.

Dirt will not work. Can not make an umbrella out of dirt. Dirt is 47% oxygen, 28% silicon, 8% aluminum, 5% iron, 3.6% calcium, 3% sodium, 3% potassium, and 2% magnesium. Air is nitrogen and oxygen. Water is hydrogen and oxygen, but no carbon.

A. J. Friedemann, *Life after Fossil Fuels*, Lecture Notes in Energy 81, https://doi.org/10.1007/978-3-030-70335-6_12

That leaves plants such as trees, crops, and grasses, also known as biomass. Biomass has both carbon and hydrogen. In decreasing order of abundance, the most common elements in biomass are carbon, oxygen, hydrogen, nitrogen, and Ca, K, Si, Mg, Al, S, Fe, P, Cl, Na, Mn, Ti. Biomass varies quite a bit in mineral composition, with carbon 35–65% of the dry weight, and hydrogen roughly 6%.

It is not surprising biomass is chemically similar to fossils fuels, which were made from plants. Mother Nature's recipe for fossil fuels is as follows: Find an anaerobic (no oxygen) basin, fill with remains of mostly marine plants, crush fields of decaying plant remains under tons of earth, pressure cook for hundreds of millions of years. Yield: One gallon of crude oil per 196,0000 tons of plants. That is a lot of plants! Imagine cruising through Kansas and having to cram 40 acres of wheat into your gas tank every 20 miles (Dukes 2003). Nature has done a whole lot of economic work for us!

Biomass Chemicals and Plastics

The IEA (2018) estimates that to produce just chemicals with biomass as feedstock and process energy (including the refining sector), rather than with natural gas, coal, or oil, would require half of the world's sustainable renewable biomass production by 2030. That much biomass would be about 2385 million metric tons of oil equivalent (MTOE) equal to 102 ExaJoules (EJ) each year. Current global primary demand for bioenergy is 1350 MTOE (57 EJ).

Bioplastics are being commercially made. But the bioplastic industry is quite small since it costs a great deal more to use biomass, and so comprises only 0.2% of the global market. Nor is this industry likely to grow, since biodegradable plastics are often made from more easily degradable fossil fuels than biomass (IEA 2018). Yet, being degradable is the main stated justification for the higher cost of bioplastics, and restricts the range of products made to mostly food-related items.

There are ideas about how biomass could be used to make chemicals, but they are far from being commercial. Liao et al. (2020) propose that wood can make four high-value industrial feedstocks: phenol, propylene, oligomers, and pulp. But more research needs to be done. The road from the laboratory to the marketplace is long and hard. First, a process must be developed in the laboratory, then a larger pilot project must succeed, and finally, the process be fully scaled up to a commercial level, where projects often fail. Cellulosic ethanol provides an object lesson. Despite promising research and scaling up to a commercial level, all US cellulosic ethanol factories have gone out of business, and so can not be considered commercial yet.

It is hard to imagine life without tires, fabric, roofing, packaging, soap, insulation, paint, plastic pipes, solvents, shoes, detergents, and toys. Vehicles are lighter and get more mileage with plastic parts that are also rustproof.

Asphalt, the black sticky bottom of a barrel of petroleum that paves our roads, is essential. Most roads are repaved with asphalt every 10 years, requiring 320 barrels per kilometer (0.62 miles). There is research being done on bioasphalt made from plant lignin, but given all the other biomass needs it is not likely enough could be found for the 2.5 million miles of US roads that would use 1.3 billion barrels of bioasphalt per decade.

Lubricants are another essential fossil fuel product that keeps all machines with moving parts running. Vegetable oils and biolubricants thus far cannot duplicate fossil lubricants and their high-temperature stability, low viscosity breakdown, oxidative resistance, length of storage, ability to operate in low temperatures, and corrosion resistance.

Recycling and Burning

It can take a 1000 years for plastic to degrade in a landfill, and over 600 years in the ocean, where 46% of the plastic consists of discarded fishing nets (Lebreton et al. 2018). Globally, over 18 trillion pounds of plastics have been produced from about 8% of the world's oil (Gibbens 2018).

So perhaps our descendants will not need to make plastics. They can reuse them. Or more likely burn them for heat and cooking, adding one more threat after energy descent. Burning plastics can release dioxin and other hazardous chemicals that can cause cancer, disrupt endocrine systems, cause birth defects, and Parkinson's disease (Biemiller 2013; Verma et al. 2016; Paddock 2019).

Recycling plastic is preferable to making plastic from biomass, especially considering the destructive aspects of farming—erosion, biodiversity loss, deforestation, pesticides, water depletion, and greenhouse gas emissions (Weiss et al. 2012; Cho 2017).

Recycling may be a no-brainer but is no simple matter. Plastics are difficult to recycle since plastics come in thousands of varieties, each with different chemistries, derivatives, and additives such as plasticizers, dyes, and chemicals.

Conclusion

It is quite likely that after fossils are gone, plastics will no longer be made since they are incredibly complex—PhDs in numerous fields make them possible—and most kinds have been around for only 50 years or less. Thwaites (2011) showed how hard replicating a complex process that we take for granted would be by performing a simple exercise: He tried to make an ordinary toaster from scratch. Even the simplest toaster had 404 parts of plastic, steel, mica, copper, and nickel. After a great deal of struggle, he was able to make the metal pieces, which mankind has made since the Iron Age. But plastics were beyond him. He would have had to refine crude oil to make propylene, which takes at least six chemical transformations to make into the simplest plastic, polyethylene.

We lived without plastics before and we can do it again. To prevent our fossil-fuel deprived descendants from burning them at home for heat and cooking, which can release polychlorinated biphenyls, mercury, halogens, heavy metals, and more, incinerators with pollution controls that can generate heat and electricity by burning plastic and other waste could be constructed. While we are at it, how about we do some serious research on recycling and converting plastics, the largest "oil spill" on earth, back into transportation fuels.

References

Biemiller A (2013) Can we safely burn used plastic objects in a domestic fireplace? No, you can't. Don't even think about it… Massachusetts Institute of Technology, School of Engineering

Cho R (2017). The truth about bioplastics. Columbia University. https://blogs.ei.columbia.edu/2017/12/13/the-truth-about-bioplastics/. Accessed 5 Nov 2020

Dukes JS (2003) Burning buried sunshine: Human consumption of ancient solar energy. Clim Chang 61:31–44

Gibbens S (2018) What you need to know about plant-based plastics. Natl Geogr https://www.nationalgeographic.com/environment/2018/11/are-bioplastics-made-from-plants-better-for-environment-ocean-plastic/. Accessed 5 Nov 2020

IEA (2018) The future of petrochemicals. Towards more sustainable plastics and fertilisers. International Energy Agency

KAUST (2020) Making more of methane. King Abdullah University of Science and Technology. News release. https://www.eurekalert.org/pub_releases/2020-09/kauo-mmo090220.php Accessed 22 Nov 2020

Lebreton L, Slat B, Ferrari F et al (2018) Evidence that the great pacific garbage patch is rapidly accumulating plastic. Sci Rep 8:4666

Liao Y, Koelewijn SF, Van den Bossche G et al (2020) A sustainable wood biorefinery for low–carbon footprint chemicals production. Science 367:1385–1390

Paddock RC (2019). To make this tofu, start by burning toxic plastic. New York Times

Thwaites T (2011) The Toaster Project: or a heroic attempt to build a simple electric appliance from scratch. Princeton Architectural Press. Hudson, NY

Verma R, Vinoda KS, Papireddy M et al (2016) Toxic pollutants from plastic waste – a review. Procedia Environ Sci 35:701–708

Weiss M, Haufe J, Carus M et al (2012) A review of the environmental impacts of biobased materials. J Ind Ecol. https://doi.org/10.1111/j.1530-9290.2012.00468.x

Chapter 13
And the Renewable Winner Is...

Not anything that is not already commercial, now that we are past or near peak oil. That rules out hydrogen, power2gas, wave, tidal, geothermal, fusion, breeder or thorium or gen IV reactors, and geothermal.

Not anything for which there are too few workable sites, such as hydropower, pumped hydro storage, and compressed air energy storage.

Nor anything finite like oil, coal-to-liquids, natural gas, or uranium.

Nor anything that is not renewable but merely rebuildable using fossil fuels, such as nuclear power, wind turbines, solar panels, or concentrated solar power since transportation, mining, and manufacturing can not be electrified.

Not anything that generates electricity, since utility-scale battery storage can not scale up.

Ding, ding, ding. We have a winner.

Biomass is the lone contender not thrown out of the arena. Biomass, your time has come back round again. The DOE estimates a billion tons of *sustainable* biomass could be harvested a year in the US (DOE 2016), though the next chapters will cast doubt on that.

Sources vary widely and include crop residues, switchgrass, forest residues, urban wood, milling, and methane from landfills. Let us not turn up our nose at biomass from wastewater and manure. Like all energy resources, biomass is distributed unevenly. Over half of current stocks exist in just 12 states, while the bottom 25 states have just 15% of biomass (Milbrandt 2005).

Biomass is the only renewable energy source that can generate the high heat needed by industry to make cement, iron, steel, aluminum, trucks, computers, electronic equipment, ceramics, bricks, and machinery.

Biodiesel is the only renewable drop-in fuel that can keep heavy-duty trucks, locomotives, and ships running.

Biomass, with a little help from hydropower, is the only renewable way to keep the electric grid up.

A. J. Friedemann, *Life after Fossil Fuels*, Lecture Notes in Energy 81, https://doi.org/10.1007/978-3-030-70335-6_13

Composted biomass is the only renewable way to replace natural gas-based fertilizer, which will also reduce the need for oil-based pesticides, herbicides, fungicides, and insecticides.

Biomass is the only feedstock candidate for over half a million products made out of fossils.

Biomass can cook food and heat homes and businesses.

Biomass, too, can and must feed the three billion people expected by 2050. Eat your vegetables!

Finally, civilizations before fossil fuels used biomass as their main energy source, so we know it works. Makes me want to hug the next tree I see.

I hear you. You have an important question you have been waiting to ask about this.

References

DOE (2016) The billion-ton report: advancing domestic resources for a thriving bioeconomy, Volume 1: Economic Availability of Feedstocks. U.S. Department of Energy, Oak Ridge National Laboratory. doi:https://doi.org/10.2172/1271651

Milbrandt A (2005) A geographic perspective on the current biomass resource availability in the U.S. National Renewable Energy Laboratory, Department of Energy

Chapter 14
Scale: How Much Biomass Is Required to Replace Fossil Fuels?

So, you ask in disbelief, after fossil fuels decline, civilization will return again to biomass?

And you wonder: Is there enough biomass to power transportation, industry, keep the electric grid up, make half a million products, replace natural gas fertilizer with compost, and heat homes and businesses for 7.8 billion people? Will there be biomass aplenty for the additional 3 billion of us tree-huggers born by 2050? You sound skeptical. Let us take a sight-seeing tour around Wood World.

Biomass, mainly wood, will be needed to build homes, buildings, fencing, boats, paneling, flooring, tables, chairs, beds, instruments, shelves, doors, window frames, implements, paper, cardboard, toys, ladders, particleboard, fiberboard, doors, barrels, desks, docks, decks, pilings, crates, and shingles…

Biomass charcoal will be needed for high heat in industry to smelt metals, fire cement, brick and ceramic kilns, melt sand into glass, and all the other goods requiring high temperatures.

In Europe, to replace only transportation oil with cellulosic ethanol would require 15 billion metric tons of plant biomass, 7.5 times more than the weight of grain and oilseed crops grown there. The volume of this bulky, low-density biomass is 200 billion cubic meters, 17 times more than the oil and coal that powers Europe—more than current diesel transport could move (Richard 2010).

Burning plants is more energy efficient than making them into biofuels. Yet even if you pulled every plant and its roots out of the ground and burned them, you would get just 94 ExaJoules (EJ), less than the 105 EJ of fossil fuels Americans burn a year (Patzek 2005). Denuded, our planet would look like Mars.

It would take a long time for the landscape to recover because photosynthesis is terribly inefficient. On average, only half a percent of sunshine is turned into biomass, and in temperate regions only during a brief growing season (NRC 2014; Smil 2015). If only plants could pack on the pounds like we do.

Fossil fuels have enormous energy densities per production area. Oil has a whopping energy density of up to 10,000 Watts (W) per square meter (m^2) since an oil

A. J. Friedemann, *Life after Fossil Fuels*, Lecture Notes in Energy 81, https://doi.org/10.1007/978-3-030-70335-6_14

well uses very little space. Huge tracts of land are required to grow corn, soybeans, and other oilseeds. When you include the area of land to grow these crops, petroleum has 40,000 times more energy density than ethanol (0.25 W/m^2), 67,000 times more than biodiesel (0.12–0.18 W/mw), and 200,000 times more than cellulosic ethanol (0.025 W/m^2) (Smil 2015). On top of that, there are large energy costs to collect, transport, process, and deliver biofuels using a massive road structure over tens of thousands of square miles.

A single pipeline can take crude oil to a refinery. No wonder we switched from biomass to fossil fuels. It takes hundreds of thousands of trucks and railcars to haul crops to biorefineries.

Of course, fossils are biofuels. But alas, not renewable, since it took over 100 million years to make them.

Every year we burn fossil fuels equal to 400 times the planet's total annual plant growth, including the microscopic plants in the ocean (Dukes 2003). And therein lies the rub. We are trying to make biofuels from plants faster than 100 million years. We need 400 more Earths.

Conclusion

Your skepticism about Wood World was well-founded. Biomass does not scale up enough to replace fossil fuels. So, we will just have to grow a lot more biomass.

What is that? You have another question. Is that possible?

References

Dukes JS (2003) Burning buried sunshine: human consumption of ancient solar energy. Clim Chang 61:31–44
NRC (2014) The Nexus of biofuels, climate change, and human health. National Academy of Sciences, Washington, DC
Patzek T (2005) The United States of America meets the planet earth. National Press Club Conference, Washington, DC
Richard T (2010) Challenges in scaling up biofuels infrastructure. Science 329:793–796
Smil V (2015) Power density: a key to understanding energy sources and uses. MIT Press, Cambridge. MA

Chapter 15
Grow More Biomass: Where Is the Land?

Biomass is needed to replace all uses of fossil fuels. For that reason, growing more plants is essential. The only other alternative would be to get rid of economic systems that depend on endless growth on a finite planet. How about a steady-state economy where Gross National Happiness was more important than GDP, with birth control, abortion, more opportunities for women, and higher taxes on children a priority until population was in balance with nature?

Well ha, that is not going to happen—who will fight for that? The New York Times reported on human population and carrying capacity only 16 times over the past 5 years, 10 times in stories that denied these were legitimate worries, and five times as one of many other problems. With a motto of "All the news that is fit to print," the newspaper has not noticed the elephant in the room.

So, dear reader, do you fancy yourself an eco-warrior and want to take on a crusade to stop the Ponzi scheme of endless growth and increasing population? If not, it is time that we begin to farm and plant trees in earnest.

For biofuels, the choice is limited to oilseed crops like soybeans for biodiesel and corn for ethanol since cellulosic ethanol is not commercial. For the rest of our needs, any kind of biomass will do for heating, cooking, keeping the electric grid up, replacing products made out of fossils (i.e., plastics), charcoal for manufacturing, and fertilizer (compost).

We are facing a fossil energy crisis at a time when we have already used half of all the biomass and forests over the past 2000 years (Schramski et al. 2015; Crowther et al. 2015). In the twentieth century, despite most of our energy use coming from fossil fuels, we used 10% of the planet's biomass (Smil 2015; Houghton et al. 2009). Massive population growth in low-income countries where more fuelwood and charcoal were consumed than ever before in history resulted in vast areas of deforestation (Smil 2017).

© The Author(s), under exclusive license to Springer Nature Switzerland AG 2021
A. J. Friedemann, *Life after Fossil Fuels*, Lecture Notes in Energy 81, https://doi.org/10.1007/978-3-030-70335-6_15

Where Is the Cropland?

Okay, a warning. Way too many statistics headed your way. But dear reader, you have shown an ability to repeatedly choke down statistics and come back for more. You are about to be subjected to what in academia is known as a "quantitative assessment."

To feed today's 7.8 billion people without fossil fuel-derived fertilizers would require half of all ice-free land on the planet instead of the 15% of land used today (Smil 2011). And where is that land? We are already affecting more than 70% of the global, ice-free land surface (IPCC 2019).

Agriculture uses half of the habitable land, though three-quarters of this terrain is unsuitable for food or biofuel crops, which is why it is used mainly for livestock (Ritchie 2017).

Yet the US is still converting precious limited cropland to other uses, with over a third lost in the last 25 years. In 1949 there were 477 million acres of cropland (Nickerson and Borchers 2012) whereas in 2015, just 366.7 million (USDA 2018), a loss of 110.3 million acres.

In the US, from 2010 to 2018 population grew by 20.7 million people—35 million births, 9.6 million legal immigrants, minus 23.9 million deaths (CDC 2010–2018; DHS 2018), not counting illegal immigration. This adds up to 2.3 million new people a year. They need food, energy, cars, homes, water, sewage, garbage disposal, schools, shopping malls, and roads. Each of those requires space.

That 2.3 million new people per year are equal to a new Houston every year. From scratch. Or a new Miami, St. Louis, Pittsburgh, Cincinnati, Cleveland, *and* Atlanta. Every year. At the rate we are growing, there will be 184 million more people in the US by 2100, nearly twice as many people as live in the 311 largest cities in the US today (USC 2018; Wiki 2017). Cities took up 68 million acres of land in 2010, and are expected to grow to 163.1 million acres by 2060 (Nowak and Greenfield 2018).

Over 85% of sprawl happens near cities, which were originally built in the locations with the best farmland and water (Kolankiewicz et al. 2014). Of the 2.43 billion acres in the US, 214 million acres are developed, 9% of the land (Nickerson and Borchers 2012). Development is especially destructive because it seals the soil with asphalt and buildings, a total loss for food and biomass, increases flooding, and decreases biodiversity, water infiltration, purification, and carbon sequestration (FAO 2015a).

Globally, urban areas expanded faster than population growth. The rate of growth of urban areas was four times than had been predicted, transforming 3750 square miles (9687 km^2) a year from 1985–2015 (Liu et al. 2020). Sing it with me: "They paved paradise, put up a parking lot."

Mark Twain said it best, "Buy land, they aren't making it anymore."

Why does this loss of farmland matter? Pimentel and Pimentel (1990) estimate that given increasing soil erosion and declining resources—food, groundwater, pollution, forests, and fossil fuels—the ideal US population at our current standard of living would be 40–100 million people. Even for me, that is hard to contemplate. Birth control anyone?

The two crops of which we would need the most are soybeans (or other oilseed crops) and corn for biofuels since without transportation everything else stops running. But these crops already take up half of the cropland, and can not necessarily be grown on the other half due to insufficient rainfall, unsuitable temperatures, and poor soil. Thank goodness though. Who wants to eat corn and soy slop day in, day out?

We could always mow down the remaining wilderness to plant soy and corn, but that will not be of much use as wilderness already is dwindling too. In just the last 25 years, 10% of the remaining global wilderness was lost. At this rate, all wilderness would be gone by 2100 (Watson et al. 2016). Perish the thought.

Much of Earth's Land Is Degraded or Unavailable

A quarter of the land on the planet is not suitable for biomass farming. It is too frozen, too wet, too dry, too rocky, too toxic, too salty, too compacted, too acidic, too steep, the soil too thin, or deficient in one or more of 16 crop nutrients. Or already developed with cities, suburbs, roads, and buildings.

Another quarter of the earth's land is degraded from erosion, unsustainable farming and grazing, pesticides, desertification, aridification, salinization, pollution, deforestation, mining, microplastics, and toxic elements from coal plants (i.e., mercury and arsenic), industries, and wars (Boots et al. 2019).

The arable land we grow food and graze animals on has degraded so much that the United Nations estimates there are only 60 years of harvests left worldwide, on average. According to the UN, 95% of our planet's land will be degraded by 2050, forcing millions to migrate as food production fails (FAO 2015b; Leahy 2018). However, another study's findings were less dire. Although 93% of soils are thinning, only 16% of the world's farms have a lifespan of less than 100 years and another third may last 100–200 years (Evans et al. 2020).

Net primary productivity (NPP) is the rate at which energy is stored as biomass by plants or other primary producers and made available to the consumers in the ecosystem. A huge percent of the earth's NPP, 40% on average, is already used by humans for food, fiber, animal feed, lumber, fuel, and grazing. Some regions use far more, with Europe at 70% and South-Central Asia at 80% of NPP (Imhoff et al. 2004).

What is left is used by millions of species. Any dramatic human increase of NPP will drive other species extinct and harm ecosystems (NRC 2014; Smil 2010). These ecosystems provide at least $33 trillion in services a year that keep us alive (Costanza et al. 1997) by purifying water, forming new soil, sequestering carbon, regulating climate, and fostering biodiversity.

Let's Grow Food and Biofuels on Other Nation's Land

This must be a winning idea. Turns out, many nations are already doing this. A land grab has been underway for many years. China, India, Saudi Arabia, and other nations have leased vast areas of Africa, Patagonia, Brazil, Paraguay, Ukraine, Indonesia, Papua New Guinea, Cambodia, and foreign investors own 30 million acres of US farmland (Pearce 2012; Wilde 2019). Any nation just now entertaining this idea is late to the game. Of course, the US is a relatively less densely populated land than most other countries. And remember, to bring the goods back to their countries, they still have to pay those shipping energy costs.

Food Production Shows Signs of Peaking

The green revolution was amazing, but its progress is stalling. The rate of increase in food production is slowing down with a peak food rate of growth in sight. This is a confusing concept because the peak rate is not at all the same as peak food production. The amount of food produced is still increasing. But the rate of increase is slowing down. Think of it like you have ridden the roller coaster to its top, and when you look ahead, you see the dreaded Thomas Malthus drop off coming up below. Some foods already have decreasing yields.

The rate of growth of 14 types of food that provide the majority of our calories had an average peak rate year of 2006. We can not just swap one for another—corn for pork, eggs for wheat, milk for fish (Seppelt et al. 2014). Yet, economists believe there is a substitution for everything so that we will never experience shortages of anything, an idea critically discussed and challenged by Hall and Klitgaard (2018). These economists who believe in the religion of substitution might as well say "let them eat cake." With food in short supply, it would instead be a cake made of grass and bark.

Genetically Engineer Plants to Grow Faster, Get Larger

Photosynthesis evolved about three billion years ago, and to this day, only converts a tiny fraction of sunlight into biomass. So seriously—we are going to enhance photosynthesis when Mother Nature did not figure that out over three billion years of random mutations? It is possible improved photosynthesis would make a plant less disease-resistant, or put more growth into leaves and stalks rather than edible fruit or grain, or require yet more water and soil nutrition. There are probably good reasons and limitations keeping nature from improving photosynthesis.

Crops have always produced carbohydrates and sugars shared with soil organisms in exchange for nitrogen, phosphorus, potassium, minerals, and disease protection. But it appears that plants bred for the green revolution, able to get these services from man-made fertilizers, may have stopped working symbiotically with soil microbes as well (Porter and Sachs 2020). So green revolution plants bred to feed an additional billion may not do as well post natural gas fertilizer.

Post-Harvest Food Loss

In the past and today in less-developed countries, 50–60% of crops may be lost after harvest from excess moisture, spoilage, varmints, diseases, mold, and fungi. Or losses can be kept as low as 1–2% through technology (Kumar and Kalita 2017). This requires gigantic half-mile long 120-foot high granaries made of steel and concrete that keep rodents, birds, disease, and insects out. Years of grains can be stored after harvest in storage bins kept at perfect temperatures and ventilation. These huge granaries are often hundreds or even thousands of miles away from the grain's destination in the US and overseas (Golob et al. 2002). After fossil fuels, much smaller wood, brick, and clay granaries may lose half or more of post-harvest crops again.

The Cool-Chain Will Be Far More Local

Today a third of America's fresh produce is grown in California and shipped thousands of miles to the world. Crops are harvested early in the morning when temperatures are under 60 F and put into air-conditioned warehouses within 2 h. If picked at temperatures over 75 F and not cooled for 10 h, spoilage starts within 2.5 days before it can reach customers.

Conclusion: Too Little Land, Too Many People

With another three billion people expected in the next 30 years, we will need to increase food production 70–100% by 2050 (FAO 2015a). Doubling yields in the next 30 years would require an annual yield increase of 2.2%, more than the average annual increase of the past 50 years (Weber and Bar-Even 2019) at a time when the rate of increase is decreasing—peak food.

As Mark Twain observed, land is a constraint. More biomass production is not likely due to a lack of arable land.

Clearly, the planet will not be able to support the present or an unchecked future population. Clearly, family planning is the best solution. Fewer humans mean more biomass for all, but family planning is a taboo topic. Right-wing religions, politicians, and capitalists would seem to prefer the Four Horsemen to arrive in the future rather than to forgo ever-increasing profits (Hardin 1993; Hardin 1999; Friedemann 2017; Beck and Kolankiewicz 2000).

Desperate times demand desperate measures. Or absurd ones. Would the powers that be want to genetically engineer humans to be able to eat plentiful grass, bark, and leaves, devolving back to the Planet of the Apes? Or how about "Honey, I shrunk the kids?" If humans were bred to be really tiny, there will be plenty to go around as well.

Simply put, we need to take care of the land we have left to grow more food. And we need to put population discussion back on the dinner table.

References

Beck R, Kolankiewicz L (2000) The environmental movement's retreat from advocating U.S. population stabilization (1970-1998): a first draft of history. J Policy Hist 12:123–156

Boots B, Russell CW, Green DS (2019) Effects of microplastics in soil ecosystems: above and below ground. Environ Sci Technol 53:11496–11506

CDC (2010–2018) National vital statistics reports: deaths and births and natality. Center for Disease Control and prevention

Costanza R, d'Arge R, de Groot R et al (1997) The value of the world's ecosystem services and natural capital. Nature 387:253–260

Crowther TW, Glick HB, Bradford MA (2015) Mapping tree density at a global scale. Nature 525:201–205

DHS (2018) Yearbook of immigration statistics 2017: Table 1. Persons obtaining lawful permanent resident status: fiscal years 1820 to 2017 & Estimates of the Illegal Alien Population Residing in the United States: January 2015 Department of Homeland Security

Evans DL, Quinton JN, Davies JAC et al (2020) Soil lifespans and how they can be extended by land use and management change. Environ Res Lett 15:0940b2

FAO (2015a) Status of the World's Soil Resources Main Report. Food and Agriculture Organization of the United Nations and Intergovernmental Technical Panel on Soils, Rome, Italy

FAO (2015b) International year of soil conference 2015. Healthy soils for a healthy life. United Nations Food & Agriculture Organization

Friedemann A (2017) Why did everyone stop talking about population and immigration? Energyskeptic. http://energyskeptic.com/2017/why-are-population-immigration-taboo-topics/. Accessed 5 Nov 2020

Golob P, Farrell G, Orchard JE (2002) Crop post-harvest: science and technology. In: Principles and practice, vol 1. Blackwell Science Ltd., Oxford

Hall CAS, Klitgaard K (2018) Energy and the wealth of nations. An introduction to biophysical economics, 2nd edn. Springer, New York, NY

Hardin G (1993) Living within limits: ecology, economics, and population taboos. Oxford University Press, New York

Hardin G (1999) The ostrich factor: our population myopia. Oxford University Press, New York

Houghton RA, Hall F, Goetz SJ (2009) Importance of biomass in the global carbon cycle. J Geophys Res 114

Imhoff ML, Bounoua L, Ricketts T et al (2004) Global patterns in human consumption of net primary production. Table 3. Nature 429(6994):870–873

IPCC (2019) IPCC special report on climate change, desertification, land degradation, sustainable land management, food security, and greenhouse gas fluxes in terrestrial ecosystems. Intergovernmental Panel on Climate Change

Kolankiewicz L, Beck R, Manetas A (2014) Vanishing open spaces population growth and sprawl in America. Numbers USA, Arlington, VA

Kumar D, Kalita P (2017) Reducing postharvest losses during storage of grain crops to strengthen food security in developing countries. Foods. https://doi.org/10.3390/foods6010008

Leahy S (2018) 75% of Earth's land areas are degraded. Natl Geogr. https://www.nationalgeographic.com/news/2018/03/ipbes-land-degradation-environmental-damage-report-spd/

Liu X, Huang Y, Xu X et al (2020) High-spatiotemporal-resolution mapping of global urban change from 1985 to 2015. Nat Sustain 3:564–570

Nickerson C, Borchers A (2012) How is land in the United States used? A focus on agricultural land. U.S. Department of Agriculture

Nowak DJ, Greenfield EJ (2018) US urban forest statistics, values, and projections. J For 116:164–177

NRC (2014) The nexus of biofuels, climate change, and human health: workshop summary. Institute of Medicine. National Research Council, National Academies Press, Washington, DC

Pearce F (2012) The land grabbers: the new fight over who owns the earth. Beacon Press, Boston, MA

Pimentel D, Pimentel M (1990) Land, energy, and water: the constraints governing ideal U.S. population size. The NPG forum. PMID: 12178968

Porter SS, Sachs JL (2020) Agriculture and the disruption of plant-microbial symbiosis. Trends Ecol Evol 35:426–439

Ritchie H (2017) How much of the world's land would we need in order to feed the global population with the average diet of a given country? https://ourworldindata.org/agricultural-land-by-global-diets

Schramski JR, Gattie DK, Brown JH (2015) Human domination of the biosphere: rapid discharge of the earth-space battery foretells the future of humankind. Proc Natl Acad Sci 112:9511–9530

Seppelt R, Manceur AM, Liu J et al (2014) Synchronized peak-rate years of global resources use. Ecol Soc 19:50

Smil V (2010) Energy transitions: history, requirements, prospects. Praeger, Santa Barbara, CA

Smil V (2011) Harvesting the biosphere: how much have we taken from nature. The MIT Press, Cambridge, MA

Smil V (2015) Power density: a key to understanding energy sources and uses. MIT Press, Cambridge, MA

Smil V (2017) Energy and civilization a history. MIT Press, Cambridge, MA

USC (2018) Table 4. Cumulative Estimates of the Components of Resident Population Change for the United States, Regions, States, and Puerto Rico: April 1, 2010 to July 1, 2018. United States Census

USDA (2018) Summary Report: 2015 National Resources Inventory. U.S. Department of Agriculture, Natural Resources Conservation Service, Washington, DC, and Center for Survey Statistics and Methodology, Iowa State University, Ames, Iowa

Watson JEM, Shanahan DF, Di Marco M et al (2016) Catastrophic declines in wilderness areas undermine global environment targets. Curr Biol 26:2929–2934

Weber APM, Bar-Even A (2019) Improving the efficiency of photosynthetic carbon reactions. Plant Physiol

Wiki (2017) List of United States cities by population. Wikipedia.org

Wilde R (2019) 'American soil' is increasingly foreign owned. NPR.org. https://www.npr.org/2019/05/27/723501793/american-soil-is-increasingly-foreign-owned. Accessed 5 Nov 2020

Chapter 16
The Ground is Disappearing Beneath Our Feet

We do not think of soil as something that moves. And yet it does. Soil moves by escaping in the wind and washing away in storms. Erosion is as old as the wind and the rain, and it happens fastest on slopes, bare soils after harvest, and crops with wide rows like corn and soybeans.

Although there will always be miles of dirt and rocks beneath our feet, the living topsoil skin where crops can grow is thin and, on the move, eroding 10–1000 times faster than in untouched nature (UNFAO 2015).

Yikes: This thin layer of soil is all that stands between us and starvation. An average of 12 inches is essential for crops to thrive (Evans et al. 2020). Dig below that and you hit death row, a subsoil desert without the rich organic matter plants dig their roots into to get the nutrients and water they need. The deeper the topsoil, ideally a foot or more, the more crops produced, and the cheaper the price to grow them (Al-Kaisi 2001).

Soil grows back very slowly. An inch of fertile, organic topsoil takes 300 to several thousand years to create depending on what kinds of minerals are below in the subsoil (Bogard 2017). The fastest rate of soil formation occurs in hot, wet areas, and the slowest in areas that are cold and dry. It takes centuries because below the soil there is rock that needs to break down into smaller pieces (SSA 2020).

Iowa has lost half its topsoil in less than 100 years, from an average of 18 to 10 inches deep (Klee 1991). Another study says it has gone from 14 to 18 inches deep to 6–8 inches (Needelman 2013). Illinois topsoil has also declined by half over a century.

But why pick on Iowa and Illinois. All states are in the erosion club and have lost about half their organic matter since colonial times (Hopkinson 2017). And not just in the US. The world has lost half its topsoil over the past 150 years (UNFAO 2015).

A. J. Friedemann, *Life after Fossil Fuels*, Lecture Notes in Energy 81, https://doi.org/10.1007/978-3-030-70335-6_16

Why is Soil Erosion Happening Much Faster Now than in the Past?

Civilizations before fossil fuels typically endured about 800–2000 years before they eroded their soil to the point where they collapsed. Recall the stories of the Fertile Crescent of the Middle East, the Greek, Roman, and Mesopotamian empires. Much of this erosion resulted from forests being cut down for farmland, cooking, heating, smelting metals, ceramics, homes, ships, and wagons. Extensive logging exposed soil to wind and water. Nor were most crop residues composted and returned to the soil for next year's crop nutrition. Instead, residues were used for animal feed, thatching roofs, mattress stuffing, or burned (Montgomery 2007; Perlin 2005).

Today industrial agriculture is on track to crash civilization in just a few centuries. Mechanized plows introduced in the 1930s accelerated erosion by plowing the upper 6–8 inches of earth, exposing the soil to rain and wind. Gigantic tractors and other farm machinery with lots of horsepower have accelerated the loss. On average, this has led to 90 times more soil lost than formed (Coombs 2007).

Farmers who rent land have no financial incentive to preserve the land for future generations. This is accentuated as family farms have evolved to corporate farms. Not surprisingly, soil erosion is greater on the 54% of cropland that is rented (NRCS 2010; NSAC 2015; Ranjan et al. 2019; USDA 2019).

Even people who farm their own land neglect erosion control because it can take decades or centuries before yields decline noticeably (UNFAO 2015). Everything seems fine until it is not.

The Great Dust Bowl of the 1930s was a wake-up call. Drought and erosion across the American and Canadian prairies affected 100,000,000 acres, devastating the panhandles of Texas and Oklahoma and scouring adjacent sections of New Mexico, Colorado, and Kansas. The phenomenon was caused by severe drought and dryland farming practices that exposed the land to wind erosion, climaxing in a period of epic dust storms.

There are many ways to slow erosion down, which the National Soil Conservation Service has been helping farmers do since the 1930s Dust Bowl. But erosion control can be expensive, and a farmer's first priority is paying off debts for machinery and farm inputs. Nor are there effective laws against erosion, nor against poisoning land, waterways, and air with pesticides and fertilizers that will affect generations for millennia into the future.

Soils take a lot of punishment. Industrial farm machinery compacts the soil so much that yields can be reduced by over 60%. Even permanently (McGarry and Sharp 2003; Drewry et al. 2008; Sidhu and Duiker 2006; Håkansson and Lipiec 2000). A farm tractor can weigh 60,000 pounds, compacting the soil and making it hard for roots to get water and air. That is 30 times more weight than a draught horse.

Windbreaks can lessen erosion. But when farmers feel compelled to farm every square inch, there is no room for trees or shrubs, which prevent soil from being blown and washed away. Windbreaks can also improve the land by increasing biodiversity and wildlife, protecting water and air quality, attracting beneficial insects for pest control and bees for pollination. They produce timber, reduce rural home energy costs, and can be planted with fruit or nut crops (Long and Anderson 2010).

Conclusion

The world's soils are degrading. By 2050, soil erosion is likely to lead to 30% less food being grown (Bogard 2017). And even greater losses of food if energy decline has begun, depriving tractors and distribution trucks of diesel.

Organic farming goes a long way toward solving the problem of degrading soils. If we want to grow more biomass to replace fossil fuels, then organic farming is well-matched to the task. The US spends about $80 billion every year on a farm bill to support agriculture. That is a lot of green stuff. Historically, farm bills have provided financial support for commodity crops (such as wheat, corn, soybeans, rice) and no financial support for fruits and vegetables. Why not have the next farm bill subsidize organic farming and erosion control instead of giving most of the subsidies and tax benefits to ethanol plants and multi-million-dollar farms growing commodity crops? More horses would reduce compaction, increase soil fertility, and prepare us for the day that energy decline makes tractors uneconomical. The planet of the horses is much more appealing than the planet of the apes don't you think?

References

Al-Kaisi M (2001) Soil erosion and crop productivity: topsoil thickness. Iowa State University

Bogard P (2017) The ground beneath us: from the oldest cities to the last wilderness, what dirt tells us about who we are. Little, Brown & Co.

Coombs A (2007) The dirty truth about plowing. Science

Drewry JJ, Cameron KC, Buchan GD (2008) Pasture yield and soil physical property responses to soil compaction from treading and grazing — a review. Soil Res 46:237–256

Evans DL, Quinton JN, Davies JAC et al (2020) Soil lifespans and how they can be extended by land use and management change. Environ Res Lett:15

Håkansson I, Lipiec J (2000) A review of the usefulness of relative bulk density values in studies of soil structure and compaction. Soil Tillage Res 53:71–85

Hopkinson J (2017) Can American soil be brought back to life? Politico. https://www.politico.com/agenda/story/2017/09/13/soil-health-agriculture-trend-usda-000513/. Accessed 5 Nov 2020

Klee G (1991) Conservation of natural resources. Prentice Hall

Long RF, Anderson JH (2010) Establishing hedgerows on farms in California. University of California, Agriculture and Natural Resources

McGarry D, Sharp G (2003) A rapid, immediate, farmer-usable method of assessing soil structure condition to support conservation agriculture. Conservation agriculture. In: García-Torres L, Benites J, Martínez-Vilela A, Holgado-Cabrera A (eds) Conservation agriculture. Springer, Dordrecht. https://doi.org/10.1007/978-94-017-1143-2_45

Montgomery DR (2007) Dirt: the Erosion of civilizations. University of California Press

Needelman BA (2013) What are soils? Nature education knowledge 4

NRCS (2010) 2007 National resources inventory. Soil erosion on cropland. Natural Resources Conservation Service

NSAC (2015) Who owns U.S. farmland, and how will it change? Figure 1. National sustainable agriculture coalition. https://sustainableagriculture.net/blog/total-2014-results/. Accessed 5 Nov 2020

Perlin J (2005) A forest journey: the story of wood and civilization. Countryman Press

Ranjan P, Wardropper CB, Eanes FR (2019) Understanding barriers and opportunities for adoption of conservation practices on rented farmland in the US. Land Use Policy 80:214–223

Sidhu D, Duiker SW (2006) Soil compaction in conservation tillage: crop impacts. Agron J 98:1257–1264

SSA (2020) What is soil? How do soils form? What are soil types? Soil Society of America. https://www.soils.org/about-soils/basics. Accessed 5 Nov 2020

UNFAO (2015) Status of the World's soil resources chapters 5 and 6. Food and Agriculture Organization of the United Nations

USDA (2019) Farmland ownership and tenure. U.S. Department of Agriculture, Economic Research Service

Chapter 17
Grow More Biomass: Phosphorus Fertilizer

Phosphorus needs a champion.

Your body is mainly oxygen, carbon, and hydrogen. Phosphorus is an important 1%, found in your bones, teeth, DNA, RNA, and the adenosine triphosphate that fuels all living cells. It is second to nitrogen as the most limiting element for plant growth on 40% of the world's arable land. With too little phosphorus, plants are stunted with low yields. With enough, crop yields can increase by 50%.

Much of the phosphorus in the soil is available only when soluble and not tightly bound with calcium, iron, and aluminum, as is typically the case (UNFAO 2015). In the past, farmers added soluble phosphorus using compost, animal manure, fish, ash, bones, guano, and human excrement. Waste not, want not. Or should I say, waste wanted?

Today we use mined rock phosphate but to little avail. Over half is lost from erosion on agricultural lands (Alewell et al. 2020), as well as from storm runoff, sewage released to waterways and landfill, and crop exports. The loss is further exacerbated by transforming 40% of corn and 31.6% of soybean crops to biofuels, a waste of phosphate and energy.

Phosphorus is Hard to Come by

Minable phosphorus is a relatively rare commodity. Production in the US has been declining 4–5% a year since about 1980. Walan et al. (2014) cite 19 studies that estimate the peak year or full depletion of phosphate rock that ranges from 30 to 400 years. The US has about 25 years left, 65% of it in Florida. And be nice to Moroccans—they have up to 75% of the remaining phosphorus reserves (USGS 2020; Stewart et al. 2005; Cordell et al. 2009). The entire continent of South America has essentially no phosphorous, but most of the increase in the world's grain and soy crop comes from there.

A. J. Friedemann, *Life after Fossil Fuels*, Lecture Notes in Energy 81, https://doi.org/10.1007/978-3-030-70335-6_17

Taking the long view, it might be time to take a very close look at manure. Ewww! Because manure is a rich source of phosphorus, we should realize that we are wasting our waste. Yes, we should be recycling human manure. Currently, very little phosphorus from "we the people" is being applied to farmland. Cost is an obstacle. Sewage treatments able to remove 95% of pathogens and eight heavy metals are expensive, and then the sludge must be transported to farms. So, most phosphorus is lost. The good news is that we have a motherlode of manure. And it is renewable!

There is no element that can substitute for phosphorus, nor can it be manufactured (Goeller and Weinberg 1978). The closest element in the same family of the periodic table is arsenic. Economists often make the case that substitutions are always available. Good luck with that economists!

Phosphorus Runoff Can Harm Ecosystems

Fertilizer runoff causes algal-cyanobacteria blooms that can release toxins causing liver, kidney, and brain damage. Ninety percent of drinking water lakes, rivers, and reservoirs are vulnerable to these toxic blooms. Because of this, in 2014 Toledo, Ohio had to shut down its water system for a week affecting half a million customers (Patel and Parshina-Kottas 2017). Toledo spent $500 million to avert future algal blooms (Hamers 2018). Most cities can not afford to do this (Penrod 2018).

This is happening all over the country. Flows of phosphorus from farms and cattle ranches into Florida's Lake Okeechobee—the nation's 10th largest freshwater lake—have fertilized the growth of horrific algae blooms that subsequently have been discharged to the ocean, fouling Florida's beaches with waves of red tide. You do not have to look far to find this problem. In my own backyard, Lake Temescal in Oakland, California, suffers from such toxic algal blooms, compelling the park to put up signs warning the public not to fish or even touch the water.

References

Alewell C, Ringeval B, Ballabio C et al (2020) Global phosphorus shortage will be aggravated by soil erosion. Nat Commun:11
Cordell D, Drangert JO, White S (2009) The story of phosphorus: global food security and food for thought. Glob Environ Chang 19:292–305
Goeller HE, Weinberg AM (1978) The age of substitutability. Am Econ Rev 68:1–11
Hamers L (2018) We're probably undervaluing healthy lakes and rivers. ScienceNews. https://www.sciencenews.org/blog/science-the-public/were-probably-undervaluing-healthy-lakes-and-rivers. Accessed 5 Nov 2020
Patel JK, Parshina-Kottas Y (2017) Miles of algae covering Lake Erie. New York Times. https://www.nytimes.com/interactive/2017/10/03/science/earth/lake-erie.html. Accessed 5 Nov 2020
Penrod E 2018 Deadly toxins in water supply a wake-up call for local U.S. authorities. Newsweek

Stewart W, Hammond LL, Van Kauwenbergh SJ (2005) Phosphorus as a natural resource. Phosphorus: Agriculture and the Environment 46. https://doi.org/10.2134/agronmonogr46.c1

UNFAO (2015) Status of the world's soil resources main report. United Nations Food & Agriculture Organization

USGS (2020) Phosphate rock. World Mine Production and Reserves. U.S. Geological Survey. https://pubs.usgs.gov/periodicals/mcs2020/mcs2020-phosphate.pdf. Accessed 5 Nov 2020

Walan P, Davidsson S, Johansson S et al (2014) Phosphate rock production and depletion: regional disaggregated modeling and global implications. Resour Conserv Recycl 93:178–187

Chapter 18
Grow More Biomass: Climate Change

Farmers have always been at the mercy of the weather. Climate change is making agriculture even more fraught, jeopardizing crops due to less dependable weather than in the past. The climate is producing bizarre events. Across the globe, we are learning to live in Bizarro World. In August 2020, a violent, fast-moving thunderstorm complex tore a 700-mile path from Nebraska to Indiana. Ripping across more than 10 million acres of Iowa's corn and soybean crop, the Washington Post reported that it damaged 43% of the state's corn and soybean crop. Meteorologists called this event a "derecho." For many of us who have never before heard this term, we now have a new entry in the pantheon of named destructive natural events. A derecho.

There are droughts, desertification, wildfires, freezes, hail storms, hurricanes, and tornadoes. Farmers must also deal with invasive pests, weeds, extreme heat, and pests surviving milder winters and developing resistance to pesticides and diseases. And they must contend with historic levels of flooding as slower and wetter moving hurricanes dump several feet of rain (Zhang et al. 2020).

Accelerating climate change, tropical forests in the Amazon and Southeast Asia have been cut down to grow soybeans and oil palms. Removing the forest cover exposes carbon-loaded peat soils with 10 times the CO_2 of other lands, releasing CO_2 and putting Brazil and Indonesia into the top 10 greenhouse gas emitting nations for over a decade (Fargione et al. 2008; Plevin et al. 2010).

Tipping Points

Extreme levels of destruction can happen rapidly when ecosystems reach an irreversible tipping point. One ecosystem at risk: The Amazon rainforest could turn into savanna or grassland from a runaway feedback loop of drought, wildfire, and logging (Staal et al. 2020; Watts 2019).

© The Author(s), under exclusive license to Springer Nature Switzerland AG 2021
A. J. Friedemann, *Life after Fossil Fuels*, Lecture Notes in Energy 81,
https://doi.org/10.1007/978-3-030-70335-6_18

Earth's dryland ecosystems, characterized by a lack of water, cover nearly half of the world. Think of the US Southwest, the Middle East, and Australia. As aridity increases, these areas can undergo abrupt changes resulting in drastic reductions in the ability of plants to fix carbon. More fires, dust bowls, greater erosion, a substantial decline of soil fertility, and crop failure can result. Ultimately even drought-tolerant plants disappear when the land turns into desert. Over a fifth of land may succumb to aridity by 2100 as the climate continues to change (Berdugo et al. 2020; Harrison 2019).

In the US, the entire Southwest may be returning to the drought state that existed for thousands of years before European settlers arrived in the nineteenth century during an unusually long 150-year wet period (Ingram and Malamud-Roam 2013; Steiger et al. 2019).

Drought and Heat

Wheat, maize, rice, barley, oats, rye, and sorghum—these are our vital cereal crops. High temperatures reduce cereal crop yields since crops grow too fast for the grain to mature fully. Heat damages pollination, enzymes, tissues, impairs flowering, and lowers photosynthesis rates. Heat causes plants to need more of everything at a faster rate—water, nutrients, and sunlight. Any deficiency, especially water, will lower yields. It has been shown that yields of corn and soybeans decline nonlinearly with temperatures above $30\,°C/86\,°F$ (Schauberger et al. 2017).

Heat, drought, and lack of rain are a triple whammy for food and biomass since plants need more water, which they transpire into the air. This can cause water shortages, so farmers tap lakes and rivers (which are also evaporating more) as well as finite aquifers.

Above $95\,°F$, photosynthesis drops off. The higher the heat, the more likely that the chemical byproducts created by photosynthesis will do damage. At $104\,°F$, plants go into thermal shock and photosynthesis stops, leading to severe and even total production losses (NRC 2014; Stokstad 2020).

Floods and Wind

As climate change plays out, some regions are expected to experience more rainfall and storms. Areas with just 2% more rainfall and wind could see a 30–274% rise in erosion by 2059 (O'Neal et al. 2005). Flooding like what happened in the Midwest in 2019 is especially destructive. The soil not only washes away but drains nutrients; and reduces oxygen the plant roots need to breathe, which can damage or kill the crop. If heavy machinery is used while the soil is wet, the soil will be compacted, limiting future root growth and oxygen for years. Floods kill the fungi that colonize the root systems of 95% of all plants on earth, a symbiotic relationship that helps plants take up nutrients. The loss of fungi, oxygen, and nutrients can also lower crop production (Ippolito and Al-Kaisi 2019).

Aridity and heat waves are exacerbated in breadbasket regions by northern jet streams that lock in place. Affected areas, where a quarter of the world's food is grown, can all have crop failures at the same time. Since food is traded globally, this may cause food shortages and price spikes even in distant nations (Kornhuber et al. 2019).

It is possible that climate change will increase the number and severity of severe wind storms, hurricanes, tornadoes, and derechos.

Forecasts Call for More Pests, More Weeds

In some regions, it will be warmer, and pests and weeds will be able to expand their range toward the Earth's poles (Bebber et al. 2013). Insects already eat 5–20% of major grain crops. The models of Deutsch et al. (2018) show an additional 10–25% loss of wheat, rice, and maize to insects per degree of Celsius warming, especially in the temperate zone.

A warmer climate will increase the need for pesticides. Today, northern states might apply 0–5 insecticide treatments a year to kill moths on sweet corn. In the future, Florida's current rate of 15–32 annual applications may become the norm (Hatfield et al. 2011). Pesticides in areas of more rain will be washed off, requiring more applications. Pests like the European corn borer that can produce only one generation a year in the corn belt may be able to reproduce two or more times as they do in the southern corn belt (Dyer 2014).

Climate Change Effects on California Agriculture

Every region will have its own climate issues, but California is especially important. California provides a quarter of America's food with 300 days of sunshine, the world's largest expanse of first-rate soil, irrigation from snowmelt and aquifers, and an ideal range of temperature swing for growing plants. Winters are cool but not freezing—a paradise for dozens of crops that can not take the hot summer weather (Bittman 2012; Sunset 2004).

California produces two-thirds of US fruits and nuts. These trees need chill hours, temperatures between 32- and 45-degrees Fahrenheit. In 2000, there were 30% fewer chill hours than in 1950. At current rates of climate change, sometime between 2050 and 2100, fruit and nut trees no longer will be able to chill out (Luedeling et al. 2009).

Banks of snow in the Sierra Nevada mountain range and a benign climate allow California to produce up to three crops in a year. As snow, water is stored in the mountains until the spring and summer, gradually melting off. But with climate change, more rain falls and less snow, and the snow melts off earlier. Unable to store and use this more sudden snowmelt, California farms may be reduced from three crops a year to just two and eventually to just one. Less sustained snowmelt also reduces hydropower, 15% of California's electricity. Biomass can step in to replace some of this dispatchable electricity, though fires and drought are likely to limit supplies (Babst et al. 2019).

Agriculture accounts for 80% of water use in California. Crop water is limited by the 40 million Californians who also need water. As the population continues to increase and more homes are built on top of prime farmland, water and crops will be reduced further.

Climate models show more severe weather events in the future. This pertains to California and poses a threat to the Sacramento-San Joaquin River delta and the delta's century-old levees that protect 5% of California's most productive farmland.

In California's Central Valley every 160 years, on average, such huge amounts of rain arrive in atmospheric rivers that as much as 10 feet of rain and snow are dumped in one event. This phenomenon is known as ARkStorm and it turns the valley into a giant lake, destroying crops and orchards, washing away topsoil and polluting the land with pesticides, manure, sewage, and hazardous waste. The next ARkStorm is likely to cost over $725 billion, with 1.5 million people evacuated, up to 25% of buildings flooded, and cattle, chickens, and other livestock drowned. It will take at least 5 years for vineyards, fruit, and nut trees to be replanted and be producing again (Porter et al. 2011). It has been 160 years since the last ARkStorm. Something to look forward to!

Meanwhile, in their quest for ever more irrigation water, Central Valley farmers are draining the aquifers. Knowingly or not, they are depleting and collapsing underground aquifers permanently, causing billions of dollars in damaged infrastructure of water canals and roads as the ground subsides. California was the last state in the West to regulate the taking of groundwater. Let us hope there is water left by 2040. That is when new laws limiting the taking of groundwater to sustainable rates finally come into play.

Conclusion

The IPCC (2014) predicts that by 2050, "the world may reach a threshold of global warming beyond which current agricultural practices can no longer support large human civilizations."

References

Babst F, Bouriaud O, Poulter B et al (2019) Twentieth century redistribution in climatic drivers of global tree growth. Sci Adv. https://doi.org/10.1126/sciadv.aat4313

Bebber DP, Ramotowski MAT, Gurr SJ (2013) Crop pests and pathogens move polewards in a warming world. Nat Clim Chang 3:985–988

Berdugo M, Delgado-Baquerizo M, Soliveres S et al (2020) Global ecosystem thresholds driven by aridity. Science 367:787–790

Bittman M (2012) Everyone eats there. New York Times. https://www.nytimes.com/2012/10/14/magazine/californias-central-valley-land-of-a-billion-vegetables.html

Deutsch CA, Tewksbury JJ, Tigchelaar M et al (2018) Increase in crop losses to insect pests in a warming climate. Science 361:916–919

Dyer A (2014) Chasing the red queen. The evolutionary race between agricultural pests and poisons. Island Press

Fargione J, Hill J, Tilman D et al (2008) Land clearing and the biofuel carbon debt. Science 319:1235–1238

Harrison S (2019) The Midwest's farms face an intense, crop-killing future. Wired.com. https://www.wired.com/story/midwest-farms-face-an-intense-crop-killing-future/. Accessed 5 Nov 2020

Hatfield JL, Boote KH, Kimball BA et al (2011) Climate impacts on agriculture: implications for crop production. Agron J 103:351–370

Ingram BL, Malamud-Roam F (2013) The west without water: what past floods, Droughts, and Other Climatic Clues Tell Us about Tomorrow. University of California Press

IPCC (2014) Climate Change 2014: Impacts, Adaptation, and Vulnerability. Part A: Global and Sectoral Aspects. International Panel on Climate Change, Cambridge University Press

Ippolito J, Al-Kaisi M (2019) The dirt on soil loss from the Midwest floods. Theconversation.com. https://theconversation.com/the-dirt-on-soil-loss-from-the-midwest-floods-114423. Accessed 5 Nov 2020

Kornhuber K, Coumou D, Vogel E et al (2019) Amplified Rossby waves enhance risk of concurrent heatwaves in major breadbasket regions. Nat Clim Change 10:48–53

Luedeling E, Zhang M, Girvetz EH (2009) Climatic changes lead to declining winter chill for fruit and nut trees in California during 1950–2099. PLoS One. https://doi.org/10.1371/journal.pone.0006166

NRC (2014) The Nexus of biofuels, climate change, & human health. National Research Council, National Academy of Sciences

O'Neal MR, Nearing MA, Vining RC et al (2005) Climate change impacts on soil erosion in Midwest United States with changes in crop management. Catena 61:165–184

Plevin RJ, O'Hare M, Jones AD et al (2010) Greenhouse gas emissions from biofuels' indirect land use change are uncertain but may be much greater than previously estimated. Environ Sci Technol 44:31–8021

Porter K, Wein A, Alpers C, et al (2011) Overview of the ArkStorm scenario. Open-File Report 2010-1312. U.S. Department of the Interior, U.S. Geological Survey

Schauberger B, Archontoulis S, Arneth A et al (2017) Consistent negative response of US crops to high temperatures in observations and crop models. Nat Commun 8:13931

Staal A, Fetzer I, Wang-Erlandsson L et al (2020) Hysteresis of tropical forests in the 21st century. Nat Commun 11:4978

Steiger NJ, Smerdon JE, Cook BI et al (2019) Oceanic and radiative forcing of medieval megadroughts in the American southwest. Sci Adv 5. https://doi.org/10.1126/sciadv.aax0087

Stokstad E (2020) Heat-protected plants offer cool surprise—greater yields. Science 368:355

Sunset (2004) Cool-season crops. Sunset magazine. https://www.sunset.com/garden/garden-basics/cool-season-crops-0. Accessed 8 Nov 2020

Watts J (2019) Amazon deforestation accelerating towards unrecoverable 'tipping point'. The Guardian. https://www.theguardian.com/world/2019/jul/25/amazonian-rainforest-near-unrecoverable-tipping-point. Accessed 8 Nov 2020

Zhang G, Murakami H, Knutson TR et al (2020) Tropical cyclone motion in a changing climate. Sci Adv 6:eaaz7610

Chapter 19
Grow More Biomass: Dwindling Groundwater

I am always amazed at the desert nations of the Middle East. Cursed by lack of water, they swap oil and gas for water-intensive food and burn prodigal quantities of hydrocarbons to desalinate seawater. Ah, America. If we were only capable of fathoming the gift of abundant groundwater our country has been given, more essential for survival than oil and gas.

Over half of Americans rely on underground aquifers for drinking water (Glennon 2002). Seventy percent of our groundwater is used to grow irrigated crops. The rest is used by livestock, aquaculture, industry, mining, and thermoelectric power plants (USGS 2018).

Two of the most important aquifers in the US are the Ogallala, beneath the Great Plains, and the second in California, the multiple aquifers beneath the Central Valley. Both are in arid regions, but they are also the nation's breadbaskets. More than half of America's food is grown in these two regions.

The Ogallala aquifer sits under eight high plains states, from Texas to South Dakota, and provides the water for a third of US crops and livestock (NT 2016). The Ogallala is vast but it is not limitless. Beneath Kansas, Oklahoma, Texas, and New Mexico, water is being drawn down at an unsustainable rate. In these areas, the aquifer may be completely depleted as soon as 2050–2070 (Katz 2016). This water will not be recharged until after the next Ice Age (Fig. 19.1 left).

To give you an idea of just how fast the Ogallala is being drained, from 2000 to 2008, as much water was taken out of the aquifer as in the previous 100 years. During that time, the aquifer provided water equal to the annual volume of 18 Colorado rivers (Little 2009). The Colorado River provides enough water for 40 million people and 5.5 million acres of farmland (Sadasivam 2019).

In Texas and Kansas, grain production peaked in 2016 and has been declining since then. By 2050, grain production in Texas could be reduced by as much as 40%. Continued depletion of the Ogallala aquifer at current levels represents a significant threat to food and water security both in the US and globally (Mrad et al. 2020).

© The Author(s), under exclusive license to Springer Nature Switzerland AG 2021
A. J. Friedemann, *Life after Fossil Fuels*, Lecture Notes in Energy 81, https://doi.org/10.1007/978-3-030-70335-6_19

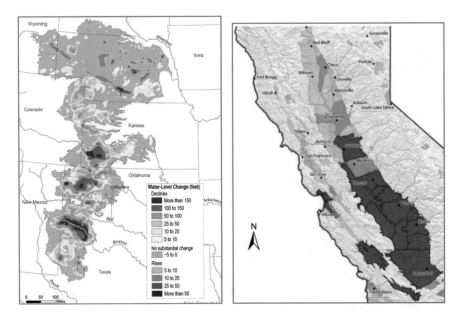

Fig. 19.1 Left: Changes in Ogallala aquifer from predevelopment to 2015 (Gowda et al. 2018). Right: Groundwater basins in California with following color key: red: critically over drafted, orange: high priority, yellow: medium priority (Henry 2019)

California's Central Valley is one of the world's most productive agricultural regions. Making up less than 1% of US farmland, the Central Valley supplies 8% of US agricultural output (by value) and produces one-quarter of the nation's food, including 40% of its fruits, nuts, and other table foods (USGS 2020). How 'bout them apples!

The Central Valley is irrigated by surface water flowing down from the Sierra Nevada mountains as well as its aquifers. These aquifers account for one-third of California's water use. They are being pumped down faster than they can be replenished, and are in decline (Fig. 19.1 right).

At the rate farmers are depleting California aquifers, which lie beneath the best soil in the nation, this region could run out of groundwater as early as the 2030s (de Graaf et al. 2015). Poof, a big bite of US food disappears from our plates. From 2000–2008, California used up a fifth of all the aquifer water that had ever existed there (Konikow 2013), and even more during the great drought of 2011–2017.

When too much groundwater is withdrawn, the ground can literally sink beneath us. Irreversible compaction can occur, causing permanent subsidence and loss of storage capacity. Subsidence also breaks roads, pipelines, and canals.

When too much water is pumped from aquifers, rivers and lakes can dry up. Saltwater may intrude, rendering water undrinkable. This problem is quite serious in California as well as Florida, Texas, and South Carolina (Glennon 2002).

Corn and Soybeans can Drink Other Crops Under the Table

Crops require a mind-boggling amount of water. It takes 13,676 liters (3613 gallons) of rainfall or irrigation water to produce enough soybeans to make just one liter (0.25 gallon) of biodiesel. Corn is more efficient, though still a heavy drinker, using 2570 L (680 gallons) of water per liter of ethanol produced (Gerbens-Leenes et al. 2009).

If corn is destined for an ethanol plant rather than your kitchen, four more gallons of water per gallon of ethanol is required. Despite 30 inches of rain per year in Iowa, that is still not enough. Unsustainable amounts of water are being pumped from aquifers to make ethanol, threatening Iowa's water supply (Harball 2013; NRC 2014). And of the top 10 corn-producing states, three are over the Ogallala: Nebraska, South Dakota, and Kansas.

In some irrigated corn acreage in the West, groundwater is being mined at a rate 25% faster than the natural recharge of its aquifer (Pimentel 2003; NRC 2011).

Conclusion

As was noted in a previous chapter, climate change, heat, and drought are reducing rainfall and surface water. Groundwater is also disappearing. It is being drained faster than aquifers can be recharged. This will only get worse as heat and drought evaporate lakes and reservoirs and plants respire more water into the sky.

Altogether, it is enough to drive you to drink!

References

de Graaf IEM, van Beek LPH, Sutanudjaja EH, et al (2015) Limits to global groundwater consumption, AGU Fall Meeting, San Francisco, California, oral presentation. https://news.agu.org/press-release/agu-fall-meeting-groundwater-resources-around-the-world-could-be-depleted-by-2050s/. Accessed 8 Nov 2020

Gerbens-Leenes W, Hoekstra AY, van der Meer TH (2009) The water footprint of bioenergy. Proc Natl Acad Sci 106:10219–10223

Glennon R (2002) Water follies. Groundwater Pumping and the Fate of America's Fresh Waters. Island Press

Gowda P, Steiner JL, Olson C et al (2018) Agriculture and rural communities. In: Reidmiller DR et al (eds) Impacts, risks, and adaptation in the united states: fourth national climate assessment, vol Volume II. U.S. Global Change Research Program, Washington, DC, pp 391–437. https://doi.org/10.7930/NCA4.2018.CH10

Harball E (2013) Rising use of corn ethanol stresses Midwestern aquifers. Underground water supplies are being pumped at an unsustainable rate thanks to corn ethanol. Scientific American

Henry L (2019) Groundwater. A firehose of paperwork is pointed at state water officials. SJV water. https://sjvwater.org/a-firehose-of-paperwork-is-pointed-at-state-water-officials/. Accessed 8 Nov 2020

Katz C (2016) As groundwater dwindles, a global food shock looms. By mid-century, says a new study, some of the biggest grain-producing regions could run dry. National Geographic

Konikow LF (2013) Groundwater depletion in the United States (1900–2008): scientific investigations report 2013–5079. U.S. Geological Survey. https://doi.org/10.3133/sir20135079

Little JB (2009) The Ogallala aquifer: saving a vital U.S. water source. Scientific American

Mrad A, Katul GG, Levia DF, et al (2020) Peak grain forecasts for the US High Plains amid withering waters. U.S. Proceedings of the National Academy of Sciences

NRC (2011) Committee on economic and environmental impacts of increasing biofuels production. National Research Council, The National Academies Press 117: 26145–26150

NRC (2014) The Nexus of biofuels, climate change, & human health. National Research Council, National Academy of Sciences

NT. 2016. Ogallala aquifer is focus of new USDA-funded research project. University of Nebraska-Lincoln.   https://news.unl.edu/newsrooms/unltoday/article/ogallala-aquifer-is-focus-of-new-usda-funded-research-project/. Accessed 8 Nov 2020

Pimentel D (2003) Ethanol fuels: energy balance, economics and environmental impacts are negative. Nat Res Res 12:127–134

Sadasivam N (2019) Will congress leave the Colorado river high and dry? Salon.com. https://www.salon.com/2019/04/06/will-congress-leave-the-colorado-river-high-and-dry_partner/. Accessed 8 Nov 2020

USGS (2018) Estimated use of water in the U.S. in 2015. Table 4A. U.S. Geological Survey

USGS (2020) California's central valley. Regional characteristics. United States Geological Survey

Chapter 20
Grow More Biomass: Vertical and Rooftop Farms

Fig. 20.1 Vertical farm. https://aerofarms.com/

As energy declines, growing food locally inside and on top of buildings is touted as a great idea. No need for diesel tractors, harvesters, and long-haul trucks and their dirty emissions. By merely pressing seeds into soil, voila! A harvest 6 weeks later using free sunshine and rain, sheltered from bad weather, plus increasing food security in the neighborhood (Fig. 20.1).

Vertical Farms in Buildings

Here is the concept: Picture walking into a library with many floors, each with many shelves, but instead of books there are thousands of plants, usually grown hydroponically with no soil, in water laced with nutrients. These are intended to be in city centers in order to grow more food locally and reduce the cost of shipping food long distances.

This is a ready road to agricultural riches? Right? Do not bet the farm on it. In a city or even a rural area, this will cost a great deal. Tall buildings, whether in New York City or Brunswick, Georgia will cost millions and millions of dollars to buy or lease space in. So, get your sugar daddy on the phone for that loan (Goodman and Minner 2019).

About that free sunlight: Somebody is unclear on the concept. Outdoors, all the leaves of a plant can be irradiated by the sun, activating photosynthesis. *That* is free sunlight. But indoors, there is simply not enough light unless it is augmented with artificial light. To equal what the sun delivers to a farm field, 100 times more light must be applied to indoor farms than is required in a typical office building (SA 2009). So, you are going to need to buy boatloads of lights.

To light up your indoor farm will take a lot of electricity. Crops like potatoes or tomatoes need about 1200 kilowatt hours (kWh) of electricity for every kilogram (2.2 pounds) of edible food produced. By comparison, an Iowa farm needs just 0.86 kWh of energy to produce two pounds of corn kernels. The average American consumes 12,000 kWh per year, so just ten pounds of potatoes will use as much electricity as a person does in a year. That means that if half of America's vegetable crops were grown in vertical farms, the lighting alone would require over half of all the electricity generated in the US (Ghanta 2010; Alternet 2016).

Moreover, researchers determined that if only solar panels were to be used to meet the energy consumption of a vertical farm, "the area of solar panels required would need to be a factor of 20 times greater than the arable area of a multi-level indoor farm" (Benke and Tomkins 2017).

Cannabis growers can shed some light on this subject. Mills (2012) used a low estimate of how much cannabis is grown indoors in the US and found this consumes about 1% of US electricity (and 3% of California's electricity) at a cost of $6 billion a year. That is equal to the energy used by about two million homes. Cannabis growers can afford to do this since marijuana commands an average price of $326 per ounce (OTC 2020) whereas fresh vegetables are worth just pennies an ounce.

Water is heavy at 8.3 pounds per gallon and needs to be pumped up to all the floors and roof. That will require a lot of electricity and a lot of water. Here are the gallons of water per pound to grow vegetables: Lettuce (15), Tomatoes (22), Cabbage (24), and potatoes (30) (Mekonnen and Hoekstra 2011; McDermott 2021).

It takes a stunning amount of energy to move water from its sources to where it is consumed. In California, the energy required to pump and move water accounts for 19% of the state's electricity usage, 30% of natural gas usage, plus 88 billion gallons of diesel fuel every year. And, demand keeps growing (Klein et al. 2005).

There is no day of rest for the vertical farmer. Fertilizer needs to be hauled to every floor, shelving erected, nutrient monitoring systems put in place, machinery acquired to harvest plants, ventilation, heating, and cooling systems installed, and shading, dehumidifiers, fans, computers, robotics, and vans to truck produce to markets purchased. Chop, chop, we have got a leak to fix!

There is only a limited range of crops that can be grown, mainly leafy greens and herbs since the better part of these plants can be consumed and require less water than other crops, which are not worth growing since they have too many inedible leaves, stems, and roots.

Sorry folks, this will not feed the neighborhood. But do not be discouraged—China is building 12-story pig farms. You could get in on the ground floor of that instead (Standaert and De Augustinis 2020).

Rooftop Gardens

Like vertical gardens, rooftop gardens save the energy required for transportation, as well as largely avoid the cost of the energy needed to heat and cool buildings.

Also, like vertical gardens, all of the soil and fertilizer will need to be hauled up and replenished, and water will need to be pumped as well (Fig. 20.2).

Unlike vertical gardens, a rooftop garden has plenty of sunlight, often too much. Even in a city with plenty of rainfall, energy-consuming water pumps and irrigation are required. Rooftop plants consume a lot of water because they get desiccated from wind speeds that can be twice as high as at ground level, and temperatures that

Fig. 20.2 Geoff Lawton at Brooklyn grange rooftop farm (Mackintosh 2014)

on average are 5 C (9 F) higher—up to 120 F from the reflected heat of other buildings—requiring them to be watered twice a day.

It is expensive to retrofit a roof to support the weight of a garden if all of the space is to be used. At best, so as not to collapse under the weight of snowfall, roofs and water-protecting membranes are built to withstand a total weight equal to 30 pounds of weight per square foot. But a cubic foot of soil weighs from 40 to 80 pounds, and water weighs about 62 pounds per cubic foot (Brady and Weil 2013), almost five times as much as the roof can bear. This is why most roof gardens are planted only around the sides, where they are supported by load-bearing walls. Crops do best in 12–24-inch-deep soil, but to spare the roof from caving in, crops that can get by in 6 inches are planted. Even though they are shallow-rooted, plants like strawberries, lettuce, beets, and radishes are not suited for rooftops since they are not heat, drought, and wind tolerant.

An aspiring rooftop farmer has to find a landlord willing to risk the potential liability, maintenance hazards, and risk of a farm failing, leaving behind tons of soil on the roof. Once a roof is secured, the headaches are not over. City codes must be met. Fire departments have regulations, and the governing board of a building may have additional rules.

Not every roof can host a garden. Buildings with small or sloped roofs will not do. Nor are buildings over 10 stories high suitable, since weather conditions are too harsh up there, and it takes too much energy to haul up soil, water, and equipment.

An ideal roof would need to be at least 10,000 square feet to be profitable for economies of scale. The building would likely be located in a manufacturing or commercial district and constructed between 1900 and 1970. During that era, building codes mandated stronger roofs able to withstand loads of up to 50 pounds per square foot. For safety, the building should never have housed heavy industries that used toxic or noxious chemicals. Mike's Asphalt Factory Farm butter lettuce is not going to be a big seller.

Farming requires a lot of space. To grow all the fruits and vegetables New Yorkers consume in a year (40.7 billion pounds), 162,000–232,000 acres of rooftops would be needed. There are about one million buildings in New York City, with 38,256 total acres of rooftop area. Buildings that meet all of the aforementioned requisite criteria reduce the total area of rooftop farms to just 3974 acres.

Rooftop farms also compete with solar panels for space.

Conclusion

Keep your feet on the ground. Sci-fi vertical and rooftop gardens are just that, science fiction, and consume more energy than farms. For all of that, they can produce very few calories.

Using vacant lots and spare land within cities to farm is a down-to-earth idea. Such urban farms may help mitigate hunger as energy declines, as long as there is enough vacant land. Detroit, for instance, has 27–40 square miles of vacant land

(17,000–25,600 acres) and just 675,000 people. That could feed a lot of people! New York City though, not so much, just 9.4 square miles (6000 acres) of green space, private, and public land for 8.6 million people, though there is green space in the suburbs and beyond.

Maybe a farm on solid ground in Motor City is the green acres place to be.

References

Alternet (2016) Why growing vegetables in high-rises is so wrong on so many levels. Alternet.org https://www.alternet.org/2016/02/why-growing-vegetables-high-rises-wrong-so-many-levels/. Accessed 8 Nov 2020

Benke K, Tomkins B (2017) Future food-production systems: vertical farming and controlled-environment agriculture. Sustain Sci Pract Policy 13:13–26

Brady NC, Weil R (2013) The nature and properties of soils, 14th edition. Pearson

Ghanta P (2010) List of foods by environmental impact and energy efficiency, Table 1. The Oil Drum. http://theoildrum.com/node/6252. Accessed 8 Nov 2020

Goodman W, Minner J. 2019. Will the urban agricultural revolution be vertical and soilless? A Case study of controlled environment agriculture in New York City. Land Use Policy 82

Klein G, Krebs M, Hall V, et al (2005) California's water-energy relationship. California Energy Commission

Mackintosh C (2014) Brooklyn grange: a rooftop farm in New York (video). Permaculture News. https://www.permaculturenews.org/2014/01/03/brooklyn-grange-rooftop-farm-new-york-video/. Accessed 8 Nov 2020

McDermott M (2021) From lettuce to beef, what's the water footprint of your food? Treehugger. https://www.treehugger.com/from-lettuce-to-beef-whats-the-water-footprint-of-your-food-4858599 Accessed Feb 5, 2021

Mekonnen MM, Hoekstra AY (2011) spikes even in distant nations of crops and derived crop products. Hydrology and Earth System Sciences 15:1577–1600

Mills E (2012) The carbon footprint of indoor *Cannabis* production. Energy Policy 46:58–67

OTC (2020) The average cost of marijuana by state. Oxford Treatment Center. https://www.oxfordtreatment.com/substance-abuse/marijuana/average-cost-of-marijuana/. Accessed 8 Nov 2020

SA (2009) Growing up: skyscraper farms seen as a way to produce food locally--and cut greenhouse emissions. Scientific American. https://www.scientificamerican.com/article/earth-talks-skyscraper-farms/. Accessed 8 Nov 2020

Standaert M, De Augustinis F (2020) A 12-storey pig farm: has China found the way to tackle animal disease? The Guardian. https://www.theguardian.com/environment/2020/sep/18/a-12-storey-pig-farm-has-china-found-a-way-to-stop-future-pandemics. Accessed 8 Nov 2020

Chapter 21
Grow More Biomass: Pesticides

Pesticides, Like Antibiotics, Are Running Out

Resistance is not futile. Just ask the makers of antibiotics and pesticides.

As microbes develop resistance, antibiotics can lose their efficacy. So too with pesticides, herbicides, fungicides, and insecticides, whether sprayed or genetically engineered into plants. Pests have fast breeding cycles, so it does not take long to develop resistance, about 5 years on average. Yet it takes on average 11.3 years and $286 million dollars to bring a new pesticide to market (McDougall 2016). The rate of discovery of new pesticides has gone almost to zero in the last 10 years or so (Borel 2017).

Before and After Pesticides

Before pesticides, when all farms were organic, farmers lost a third of their crops to pests and diseases (Dyer 2014). Synthetic chemicals came to the farm in the 1940s. And yet, a third of preharvest crops are still lost to pests (UNFAO 2015). Worse yet, 2 million tons of toxic chemicals are used by agriculture worldwide every year to prevent losses (Alavanja 2009, Sharma et al. 2019), a burden that has overflowed across our lands, sea, and air, damaging the biota that encounter them.

Farmers used to protect crops by growing dozens of kinds and rotating crop locations to confound pests. Farmers built healthy soil with compost and cover crops that contained disease-fighting microbes. Undeveloped land nearby supported bats, birds, and insects that feed on pests.

Today's farms are hundreds of square miles of wall-to-wall monoculture, single crops, a giant billboard for attracting insects that says EAT ME!

A. J. Friedemann, *Life after Fossil Fuels*, Lecture Notes in Energy 81, https://doi.org/10.1007/978-3-030-70335-6_21

The inevitability of the evolution of pesticide resistance by insect pests, with their short life cycle, has been known since 1951 when Dr. Reginald Painter at Kansas State University published "Insect Resistance in Crop Plants." Painter made a case that it would be better to understand how a crop plant fought off insects since it was inevitable that insects would develop genetic or behavioral resistance. He advocated growing many kinds of crops to enhance their innate resistance to pests instead of focusing solely on chemicals to battle pests in monocrops.

Agricultural chemicals are also killing honeybees—across Europe, bees have been found with 57 different kinds of pesticides in them (Kiljanek et al. 2016). Bees pollinate a third of what we eat, and 84% of crops depend on them and other pollinators, including fruits, vegetables, nuts, seeds, coffee, tea, and chocolate crops (Benjamin 2015).

Wild bees can be more efficient and increase yields in many crops relative to imported honeybees, especially those native to North America, and are estimated to contribute at least 20% of crop pollination. But they are dying off too, not only from pesticides but from a 200% decline in the grasslands and pastures that sustained them that were converted to corn and soy crops. If this trend continues, it may increase costs for US farmers and may even destabilize crop production over time (Losey and Vaughan 2006, Koh et al. 2015). Declining honeybee and wild bee populations are already reducing the food production of several crops (Reilly et al. 2020).

Global warming will allow weeds, insects, and diseases to thrive and expand their range, surviving winters that used to kill them, and provide a lot more time to reproduce. With the exception of the tropics, warmer temperatures will increase insect metabolic rate, causing them to eat more, and increase their reproductive rate, leading to more crop damage including predation on maize, rice, and wheat, which account for 42% of calories eaten (Deutsch et al. 2018; Lehmann et al. 2020; Ngumbi 2020).

In the case of some farms and some crops, farmers are down to just one pesticide or herbicide that still works (Borel 2017). Palmer amaranth is called the king of weeds, uniquely difficult to kill or keep from spreading (Bomgardner 2019). Each plant can make a million seeds, grows 3 inches a day up to 10 ft tall, and drastically reduces corn yields up to 91% and soybeans by 79%. The weed has evolved to resist multiple herbicides and likely requires brand-new technology that does not exist yet to control it. Palmer amaranth is not a biofuel candidate!

It's a War Out There

Before planting, farmers use fumigants to kill fungi and nematodes that survived the winter, soak the land with pre-emergent chemicals to prevent weed seeds from germinating, and then spray herbicides like glyphosate to kill any weeds that do grow. As the crop grows, several more rounds of different pesticides are applied, up to 12–30 times on cotton crops.

So many insects are being killed with insecticides that it threatens the food web that sustains life, creatures that include mammals, fish, birds, amphibians, reptiles, and even beneficial insects. It is likely that pesticides, especially neurotoxic insecticides, known as neonics, are largely responsible for the insect apocalypse that has caused terrestrial insect abundance to decline about 9% a decade (Cardoso et al. 2020, van Klink et al. 2020).

DiBartolomeis (2019) assessed the Acute Insecticide Toxicity Loading (AITL) on US agricultural land and surrounding areas and found that there was a 48-fold increase in AITL from 1992 to 2014, with corn and soybeans the most responsible. Neonicotinoids are responsible for 92% of this toxicity surge. They were introduced 25 years ago, are used on 140 crops, and do not break down for years, continuing to kill.

Obviously, our system for ensuring that pesticides are safe is flawed. It is backward, innocent until proven guilty. Rather than proving something is safe, a chemical has to be proven unsafe. Close relationships between industry and regulatory agencies help keep chemicals from being identified as toxic. By the time scientists find enough evidence of harm, the damage has already been done (Conis 2019).

Conclusion

The arms race between pesticide manufacturers and insects is escalating. For now, we have a standoff. One price is that we are poisoning our environment. With only 1% of farmland certified organically, and the end of pesticides inevitable—whether due to pest resistance or fossil fuel depletion—growing more biomass with ever more pesticides is not an option. Fossil fuels are often the feedstock for pesticides. The energy required to make these chemicals can be two to five times more than used to make the equivalent weight of nitrogen fertilizer.

The chemical war may backfire. Consider the German cockroach, first killed by chlordane in 1948. Within 3 years, cockroaches were resistant, and by 1966 they had evolved resistance to malathion, diazinon, fenthion, and DDT (Dunn 2018).

Resistance is not futile. For insects and pests, it is a way of life.

References

Alavanja MCR (2009) Pesticides use and exposure extensive worldwide. Rev Env Health 24:303–309

Benjamin A (2015) Why are bees important? You asked Google – here's the answer. The Guardian. https://www.theguardian.com/commentisfree/2015/jun/17/why-are-bees-important. Accessed 24 Nov 2020

Bomgardner MM (2019) Palmer amaranth, the king of weeds, cripples new herbicides. Scientists in the US sound the alarm about a crop-smothering weed that is growing resistant to multiple herbicides. C&EN 97. https://cen.acs.org/business/specialty-chemicals/Palmer-amaranth-king-weeds-cripples/97/i3. Accessed 9 Nov 2020

Borel B (2017) CRISPR, microbes and more are joining the war against crop killers. Nature 543:302–304

Cardoso P, Barton PS, Birkhofer K et al (2020) Scientists' warning to humanity on insect extinctions. Biol Conserv 242:108426

Conis E (2019) Why both major political parties have failed to curb dangerous pesticides. Washington Post. https://www.washingtonpost.com/outlook/2019/04/09/why-both-parties-have-failed-curb-dangerous-pesticides/. Accessed 9 Nov 2020

Deutsch CA, Tewksbury JJ, Tigchelaar M et al (2018) Insect metabolic and population growth rates predict increasing crop losses in a warming climate. Science 361:916–919

DiBartolomeis M, Kegley S, Mineau P, et al (2019) An assessment of acute insecticide toxicity loading (AITL) of chemical pesticides used on agricultural land in the United States. PLOS ONE

Dunn R (2018) Our attempts to eradicate insects are just making them resistant to pesticides. Discover. https://www.discovermagazine.com/planet-earth/our-attempts-to-eradicate-insects-are-just-making-them-resistant-to. Accessed 9 Nov 2020

Dyer A (2014) Chasing the red queen: the evolutionary race between agricultural pests and poisons. Island Press

Kiljanek T, Niewiadowska A, Semeniuk S et al (2016) Multi-residue method for the determination of pesticides and pesticide metabolites in honeybees by liquid and gas chromatography coupled with tandem mass spectrometry—honeybee poisoning incidents. J Chromatogr A 1435:100–114

Koh I, Lonsdorf EV, Williams NM, et al (2015) Modeling the status, trends, and impacts of wild bee abundance in the United States. U.S. Proceedings of the National Academy of Sciences

Losey JE, Vaughan M (2006) The economic value of ecological services provided by insects. Bioscience 56(4):311–323

McDougall P (2016) The cost of new agrochemical product discovery, development and registration in 1995, 2000, 2005–8 and 2010–2014. CropLife America and the European Crop Protection Association

Ngumbi EN (2020) How changes in weather patterns could lead to more insect invasions. The Conversation. https://theconversation.com/how-changes-in-weather-patterns-could-lead-to-more-insect-invasions-131917. Accessed 9 Nov 2020

Philipp Lehmann, et al (2020) Complex responses of global insect pests to climate warming. Frontiers in Ecology and the Environment 18 (3):141–150

Reilly JR, Artz DR, Biddinger D, et al (2020) Crop production in the USA is frequently limited by a lack of pollinators. The Royal Society

Sharma A, Kumar V, Shahzad B et al (2019) Worldwide pesticide usage and its impacts on ecosystem. SN Appl Sci 1:1446

UNFAO (2015) Keeping plant pests and diseases at bay: experts focus on global measures. United Nations Food and Agriculture Organization. http://www.fao.org/news/story/en/item/280489/icode/. Accessed 9 Nov 2020

van Klink R, Bowler DE, Gongalsky KB et al (2020) Meta-analysis reveals declines in terrestrial but increases in freshwater insect abundances. Science 368:417–420

Chapter 22
Ethanol and Energy Return on Investment (EROI)

Making fuel from plants was supposed to be an environmental triumph. We would reduce our use of crude oil and sustain oil reserves. And it would be green, carbon neutral, recycling carbon dioxide between plants and the air rather than just pumping more of it into the atmosphere.

In 2019, the US produced 15.8 billion gallons of ethanol for fuel. Almost all of it from corn. Gasoline contains 10% ethanol. This leaves 10% more oil for the future while reducing CO_2 emissions to the atmosphere, right? Sounds good! Maybe too good …

Let us take a closer look at the case for ethanol. Does it really reduce our reliance on oil? Does it make economic and environmental sense? Let us start by examining how much energy it takes to produce ethanol. Probably the best way to do this is to use the metric of EROI or Energy Return on Investment. This is simply the amount of energy delivered from some source compared to (divided by) the amount of energy required to get it. Thus, ethanol needs to be considered in light of energy return on energy invested.

If it takes one unit of fossil energy to create one unit of biofuel energy, then what is the point? All you have done is to trade the use of fossil fuel in agriculture for using the same amount in your car, with no net effect except additional soil erosion. Nothing is gained. The more steps required; the more energy required. We need to extract more energy out of a process than is required by the process itself (Hall et al. 2009). A good bit more, as Hall (2011) shows in this example of what EROI means:

- If you use twice as much energy to obtain oil as the energy in the oil can produce (EROI = 0.5:1), then you have a negative energy return of −0.5 and a lot of explaining to do.
- If you used one unit of oil to get one unit of oil (EROI = 1:1), you can stare at the hole you just made in the ground.

A. J. Friedemann, *Life after Fossil Fuels*, Lecture Notes in Energy 81, https://doi.org/10.1007/978-3-030-70335-6_22

- If you have got a slightly positive EROI of 1.1 (1.1 energy unit produced per 1 energy unit used to create it), you can pump the oil out of the ground and look at it.
- If you have got an EROI of 1.2, you can also refine the oil and look at it.
- At 1.3, you can move the oil somewhere else and look at it.

Now that you have the oil, let us consider the additional energy it will take to actually use that oil.

- If you want to run a truck on the oil you have drilled, you must add up the energy it took to build and maintain the truck, and the roads and bridges it travels on. To do that, you would need an EROI of at least 3:1
- If you want to put grain in the truck and deliver it, you need to add on the EROI of grain, bringing you to an EROI of 5:1 (This and the subsequent numbers below are less precise estimates)
- Add in the energy needed by all the workers plus the farmers, truck drivers, and their families and you are up to 7 or 8:1
- If the worker's children are to be educated, then you need an EROI of 8:1 or 9:1
- For health care 12:1
- For the arts 14:1

That may be tough to get your head around. But the concept is this simple: The energy required for almost anything in our modern lives adds up very quickly. Researchers studying the EROI of a given technology—photovoltaics for instance—often disagree about what things should be included and what should not (Prieto and Hall 2013). These boundaries can be argued over, and according to Hall (2017) may be much less divergent than some papers suggest. But this much is certain: We must be more mindful of these hidden energy costs.

Several scientists who have studied energy return ratios have estimated our civilization needs an EROI of at least 11 (Murphy 2014) or even higher, perhaps 12–14 (Lambert et al. 2014). Mearns (2008) calculated that when the net energy provided to society has an EROI of nine or less, the energy returned declines so exponentially it is like going off a cliff.

The EROI of oil is high (100:1 for some early oil fields that were easy to tap) because Mother Nature made it for free over a span of 100 million years and oil has the highest energy density of any fuel but uranium. For a long time, EROI for oil and gas in the US has been about 30:1, although it has declined to perhaps half that more recently. To make biofuels, many steps are required. Each step requires fossil fuel energy, which subtracts from the energy of the ethanol delivered.

Why Is Ethanol EROI so Low?

Consider the typical step-by-step process required to make dry-milled ethanol. Below I have described the process and identified where energy is involved as follows:

- Fossil fuel energy inputs are shown as energy-using actions in *ITALICIZED BOLD CAPITAL* letters.
- Underlined bold words are **objects made with energy**.

Before the corn gets to the ethanol factory, first you must grow the hybrid or GMO corn seed: (1) *PLANTED* with a **tractor**; (2) fields are *DOUSED* with **fertilizer** and **herbicides**; (3) *PLOWED* or *CUT* (no-till) with a **tractor;** (4) irrigated with *PUMPED* water; (5) *SPRAYED* with **pesticides**; (6) **fertilizer** *APPLIED*; (7) *HARVESTED* with a **combine**; (8) dirt and stones *REMOVED*; (9) corn kernels *SEPARATED* from the cob; (10) *DRIVEN* by **trucks** and **rail** to storage areas to sell for next year's corn ethanol crop; and (11) The hybrid or GMO corn seed is *DELIVERED* to farmers to grow corn for ethanol.

Now, you are ready to grow your corn ethanol crop. Return to step one and repeat the first 10 prior steps. With 20 steps completed, you are now at step number (21) wherein the harvested corn grain is *DRIVEN* to a **biorefinery**; (22) *UNLOADED* to a **storage area**; (23) *PLACED* on a **conveyer belt**; (24) *POURED* into a **hopper**; (25) *GROUND* into coarse flower by a **hammer mill**; (26) *POURED* onto **mesh** to select fine particles; (27) *MIXED* with *HOT* water in the **slurry mixer**; (28) *MOVED* to a **tank** and *HEATED*; (29) *MOVED* to a **jet cooker** and *HEATED*; (30) *MOVED* to a **holding column** for 10 min; (31) *MOVED* to a **liquefaction tank**; (32) *MOVED* into **fermentation tanks** to create 12–18% alcoholic "beer"; (33) *TRANSFERRED* to a **beer well tank** where settled out byproduct solids will be: (34) *CENTRIFUGED*; (35) *EVAPORATED*; (36) *DRIED*; and (37) *MOVED* to a **storage area**.

I thought you deserved a paragraph break.

Back to work. Resuming our manufacture of ethanol, proceed to (38) *LOADED* to a truck; (39) *DRIVEN* by trucks to feedlots. The liquid ethanol is then 40) *MOVED* to a **degassing tank** to remove CO_2 and other gases; (41) *MOVED* to **beer column tank**; (42) *MOVED* to the **rectifier tank** to get rid of water in a **molecular sieve column**; (43) 5% gasoline *ADDED*; (44) wastewater *TREATED (mercifully for the reader, many steps not shown)*; (45) ethanol *PUMPED* into **rail cars** or **truck tanks**; (46) **DRIVEN** by **truck** or **rail** to a **refinery** to be blended with petroleum; and (47) *DELIVERED* to **service stations**.

Rube Goldberg, do not you wish you could have lived to witness that!

You may be drained of energy by having read through all of that. No wonder! Each of these actions and objects subtract from the final EROI of ethanol. On top of that, none of these steps use ethanol as an energy source. It is fossils or 62% fossil-generated electricity all the way. How could ethanol possibly make itself after fossils are gone?

There have been dozens of studies of the EROI of corn ethanol. The best peer-reviewed research falls between a range of slightly positive to negative −0.82. So basically, break-even at best and not worth making (Giampietro et al. 1997; Pimentel and Patzek 2005; Murphy et al. 2011).

The Main Difference Between a Negative and Positive EROI Is Byproduct

Charlie Hall wondered why the results from two groups of scientists differed by so much (0.8:1 vs. 1.8:1). So, he got them to write a paper with him, Hall acting as referee (Hall et al. 2011).

In corn ethanol production, for every bushel of corn, two gallons of water, and yeast that enter the process, 36 pounds of wastes come out. A bushel yields 18 pounds of ethanol, 18 pounds of byproduct, and 18 pounds of CO_2. The corn solids byproduct, also known as dried distillers' grain (DDG), is the leftover from the ethanol-making process. Rather than discarding the DDGs as waste, they are typically used as a protein-rich animal feed. This would seem to be a good idea. Yet, the studies showed that a higher EROI resulted when the residual was not used as animal feed. That is because of the energy *not spent* producing and delivering the animal feed. Another difference: Researchers that came up with a lower EROI used larger boundaries, including, for example, the energy to produce the tractor fuel.

With DDG being fed to animals, the EROI of corn ethanol production result can be as high as 1.73 (Kim and Dale 2005; Farrell et al. 2006). That is okay if your second car is a horse and you live close to an ethanol plant.

So how come DDG is an energy waste? Scientists who study ethanol EROI explain that this byproduct is an energy loser because if left wet it takes too much energy to transport, so can only be delivered nearby, and has a short shelf life. Transporting it to distant locales means it has to be dried to reduce its weight. This further subtracts from ethanol EROI because the DDG must be centrifuged, evaporated, and transported from corn to cattle states with diesel transport. Since this corn byproduct will not be added back to the soil to replenish it, then more fertilizer will be required. Fertilizer does not grow on trees. It requires energy to make it. DDG: Waste or wasteful, take your pick!

Not included in calculating energy loss in the ethanol process is the energy to remediate topsoil erosion, aquifer depletion, biodiversity loss, and cost of water treatment plants to treat polluted farm runoff.

Christopher Portier, director of the National Center for Environmental Health asks, "If it takes one unit of fossil energy to produce 1.01 units of ethanol energy, then why are we doing this? This is not just a question of economics—it is also a health concern. If producing biofuels is not energy efficient, then the energy that is spent to make the biofuel is a pollution source contributing to health concerns." (NRC 2014).

Falling Over the Burrito Energy Cliff

The US Department of Energy has made energy easier to understand with its burrito energy unit. Really! A burrito energy unit! Why it is not a hot dog energy unit continues to perplex me.

A burrito energy unit is the amount of energy contained in one burrito. So, if a large burrito is 1200 kilocalories (kcal), then the average American needs 600 burritos a year to stay alive. Fewer burritos and the lack of calories pushes people over the energy cliff and into the grave.

Besides food, we consume energy for transportation and at home. Add up food, transportation, and home energy use employing our burrito energy conversion calculator and voila, the typical American consumes 31,000 burritos a year (DOE 2014). Feeling rich, or more likely, nauseated?

To stretch a bad metaphor past the breaking point, a barrel of oil is 5,800,000 BTU (1,462,554 kcal), so every barrel of oil is equal to 1219 burritos. Globally 92,649,000 barrels of oil are produced a day—that is 113 trillion burritos—to feed people and the rest of the world's energy needs in transportation, manufacturing, agriculture, electricity, heating, cooling, and more. A year of oil equals the energy of 41,245,000,000,000,000 burritos, and that does not even include the burritos we would have if coal and natural gas use were included. No wonder renewable energy may never catch up.

Conclusion

Ethanol is throwing good energy after bad. Farmers—correction, make that Big Ag—are making a lot of money because the process is heavily subsidized. But it is costing us dearly. Ethanol is pushing civilization closer to the edge as we use premium fossil fuels to produce relatively lower grade alcohol. At least when we fall off the energy cliff, it will be one hell of a party. Ethanol is 100% pure alcohol.

And if the concept of EROI still makes no sense to you, I have one last way of demonstrating it. Please send me $10 and I will send you back a one-dollar bill.

References

DOE (2014) How much do you consume? United States Department of Energy. https://www.energy.gov/articles/how-much-do-you-consume. Accessed 9 Nov 2020

Farrell A, Plevin RJ, Turner BT et al (2006) Ethanol can contribute to energy and environmental goals. Science 311:506–508

Giampietro M, Ulgiati S, Pimentel D (1997) Feasibility of large-scale biofuel production. Bioscience 47:587–600

Hall CAS (2011) Introduction to special issue on New Studies in EROI. (Energy Returned on Investment). Sustainability 3:1773–1777

Hall CAS (2017) Will EROI be the primary determinant of our economic future? The view of the natural scientist vs the economist. Joule 1:635–638

Hall CAS, Balogh S, Murphy DJR (2009) What is the Minimum EROI that a Sustainable Society Must Have? Energies 2:25–47

Hall CAS, Dale BE, Pimentel D (2011) Seeking to understand the reasons for different energy return on investment (EROI) estimates for biofuels. Sustainability 3:2413–2432

Kim S, Dale BE (2005) Life cycle assessment of various cropping systems utilized for producing biofuels: bioethanol and biodiesel. Biomass Bioenergy 29:426–439

Lambert JG, Hall CAS, Balogh S et al (2014) Energy, EROI and quality of life. Energy Policy 64:153–167

Mearns E (2008) The global energy crisis and its role in the pending collapse of the global economy. The oil drum. http://theoildrum.com/node/4712. Accessed 9 Nov 2020

Murphy DJ (2014) The implications of the declining energy return on investment of oil production. Philos Trans A Math Phys Eng Sci. https://doi.org/10.1098/rsta.2013.0126

Murphy DJ, Hall CAS, Powers B (2011) New perspectives on the energy return on (energy) investment (EROI) of corn ethanol. Environ Dev Sustain 13:179–202

NRC (2014) The nexus of biofuels, climate change, and human health: Workshop summary. Institute of Medicine, National Research Council, National Academies Press

Pimentel D, Patzek T (2005) Ethanol production using corn, switchgrass, and wood; biodiesel production using soybean and sunflower. Nat Resour Res 14:65–76

Prieto P, Hall CAS (2013) Spain's photovoltaic revolution: the energy return on investment. Springer, New York

Chapter 23
Corn and Soy Are Supervillains

Big Corn and Big Soy, I am going to soil your reputation. Corn and soy, the two superstars of food and biofuels, together erode more topsoil, cause more pollution, global warming, acidification, eutrophication of water, water treatment costs, fish kills, and biodiversity loss than most other crops (Powers 2005; Troeh and Thompson 2005; Zattara and Aizen 2019).

Food Versus Fuel

Over 40% of the corn crop is turned into fuel rather than food. This at a time when 43 million Americans need help with food stamps (USDA 2020) and the high unemployment rate from Covid-19 could drive the need for food aid up to over 54 million people (Lee 2020).

Too Many Pesticides

Corn and soy are especially destructive because they need a lot of pesticides. Of all pesticide use on crops, corn's share is 39.5% and soybeans 22% (Mclaughlin and Walsh 1998; Padgitt et al. 2000; Pimentel 2003; Patzek 2004; Fernandez-Cornejo et al. 2014). I do not want to say they have a drinking problem, but shall we say they have a "dependency problem." All these pesticides kill bees, wild bees, and other important pollinators. The neonic pesticides mentioned earlier that are 48 times more toxic to insect life than other chemicals are mainly used on corn and soybeans (DiBartolomeis et al. 2019).

© The Author(s), under exclusive license to Springer Nature Switzerland AG 2021
A. J. Friedemann, *Life after Fossil Fuels*, Lecture Notes in Energy 81,
https://doi.org/10.1007/978-3-030-70335-6_23

Corn and Soy Already Take Up Half of US Cropland

Corn and soy are grown on over half of America's 324 million acres of cropland (USDA 2018).

Over half!

Corn can yield 500 gallons of ethanol per acre (NRC 2014; USDA 2019). That sounds like a veritable gusher. Yet corn fuel is small potatoes. Despite a doubling of corn acreage due to the 2007 federal renewable fuel standard, the 40% of corn grown to make ethanol is a measly 10% of our US gasoline mix. In the case of diesel, 99% of what we use is petroleum diesel, and 1% is biodiesel. So even if all 324 million acres of American farmland were planted in corn and soybeans, they would barely make a dent in transportation fuels while driving food and feed prices higher.

Corn and Soy Cause the Most Soil Erosion

Corn and soy are 50 or more times more prone to soil erosion than sod crops like wheat, barley, rye, and oats. Why is that? It is because they are planted in rows much wider than other crops, up to 30 inches wide, a major highway for wind and water to barrel along and take topsoil with them (Al-Kaisi 2000; Sullivan 2004). This is exacerbated by heavy harvesting equipment that compacts and pulverizes soil into a fine powder that is more easily eroded and blown or washed away (RCN 2011; Mathews 2014).

Nevertheless, a corn ethanol gold rush is on. Farmers converted ten million acres of grassland, shrubland, wetland, and forestland into cropland between 2008 and 2016, with 2.9 million acres for corn and 2.6 million acres for soy (Lark et al. 2018).

Conservation Reserve Program (CRP) lands are protected because they retain water, support pest predators, sequester carbon, and sustain wildlife. CRP land is highly erodible if farmed. In fact, the government pays farmers not to grow crops on this land. When ethanol subsidies or corn prices are high, CRP land is often converted to corn crops. In 2007, 36.7 million acres were enrolled in the CRP program, today it is just 21.9 million acres, a loss of nearly 15 million acres.

Using CRP and undeveloped land to grow corn erases the carbon benefits of using ethanol over gasoline (Uri 2000; Tomson 2007; Searchinger et al. 2008; Fargione et al. 2008; Piñeiro et al. 2009). After the harvest, most farmers leave their soil bare, except for a minority who plant cover crops or leave corn stover on the ground. This naked soil lies unprotected from wind and heavy rain that grab soil, sediment, pesticides, and fertilizer, running away with them.

A lot of soil is lost—20–40 pounds per gallon of ethanol according to Jerald L. Schnoor, professor of civil and environmental engineering at the University of Iowa (NRC 2014).

Since 16 billion gallons of ethanol are produced per year, that is, 160–320 million tons of topsoil lost. An acre of topsoil 6–7 inches deep weighs 1000 tons, so if the soil were lost in just one area, as happened in the massive Midwestern floods of 2019 (Philpott 2019; Ippolito and Al-Kaisi 2019), 250–500 square miles of topsoil would be strip mined to the subsoil bones below. But such disastrous floods are uncommon. Usually, a fraction of an inch is lost across the 127,800 square miles planted in corn, such a small amount we do not notice. But year by year, erosion adds up, subtracting from the land. As detailed in Chap. 16, topsoil is eroding all over the world, and affects up to half of America's agricultural soil, a peril to future food and the environment (Pimentel 2006).

Corn and Soybeans Are Water Hogs (Sorry Pigs)

As readers of this book who have a photographic memory recall, it takes some 3600 gallons of water to produce enough soybeans to make a quarter gallon of biodiesel, and 680 gallons of water per liter of ethanol (Gerbens-Leenes et al. 2009). You knew that, right? Afterward, for every gallon of ethanol produced, 12 gallons of noxious sewage effluent are released that need to be treated (Schulz 2007).

Corn ethanol and soy biodiesel are not good options for the arid states of the West. Nor for California. To make just 20% of the 16 billion gallons of ethanol produced a year in the US in California would require over 8 trillion gallons out of the 8.4 trillion gallons of irrigation water now used to grow over 400 kinds of crops. Soybeans would need more water than California has available (Maupin et al. 2014). As it is, field crops like corn, soy, and cotton are draining California's aquifers more than water-intensive alfalfa, truck crops, and fruit and nut crops (Levy et al. 2020). For generations now in California, there have been fights over water between agriculture, cities, the environment, and fisheries (Fingerman et al. 2008). Thank you kindly, but please do not plant your biofuel plantations in my home state of California!

Corn's dirty secret—that corn ethanol is not a public good—is well-known. Many papers have shown that it takes about one calorie of fossil fuel to make a calorie of ethanol (e.g., Pimentel 2003, Murphy et al. 2011). This is known even in the halls of the US Congress, which has created a pork barrel for corn and soy farmers. Both the House and Senate have tried many times to repeal or reduce the amount of ethanol called for in the Renewable Fuel Standard. Yet the federal mandate that US transportation fuels have a minimum volume of biofuel remains. In 2020, the mandate is 11.56% biofuels by volume. Here is a list of just a few failed congressional reform bills: HR 424 (2011), S 1584 (2015), HR703 (2015), HR 119 (2017), HR 1314 (2017), HR 104 (2019), and HR 3427 (2019) (Source: https://crsreports.congress.gov/product/pdf/IN/IN11353).

More Fertilizer, More Dead Zones

Corn also uses more nitrogen fertilizer than most crops (Padgitt et al. 2000; Pimentel 2003; NRC 2003), and significant amounts of phosphorus. Corn needs a lot of fertilizer because corn plants are quite adept at absorbing nitrogen and storing it in the corn grain. But unfortunately, much of the nitrogen fertilizer applied does not go into the grain but instead washes away into lakes, rivers, and the ocean (NRC 2014).

Fertilizer runoff is also the main culprit causing dead zones such as the 6000–7000 square miles of water at the mouth of the Mississippi River in the Gulf of Mexico. Dead zones develop when overloads of nutrients—nitrogen and phosphorus—allow algae to go on a nutrient binge, proliferating to the point they suck the oxygen out of the water. Any fish, shrimp, and crabs in this zone die, especially bottom-dwelling fish, shrimp, clams, mussels, and oysters. Those that live can accumulate algal toxins that concentrate in shellfish, herring, mackerel, and sardines near the bottom of the food chain making them potentially lethal.

Industrial Farming Is Great for Jellyfish

Not being harmed, and in fact ballooning in numbers, are jellyfish. They are on the way to dominating the ocean and displacing fish. Despite the fact that they have no backbones! Jellyfish can handle hypoxia (low oxygen levels), love warmer climate-changed oceans, and proliferate; thanks to trawling, overfishing, fertilizer, and sewage runoff. We are tipping the ocean ecosystem to favor jellyfish, possibly permanently. And they are awfully hard to kill. Chemical repellents, biocides, nets, electric shocks, and introducing species that eat jellyfish will not do it. If you shoot, stab, slash, or chop off part of a jellyfish, it can regenerate lost body parts within 2 days. Not even the past five major extinction events that killed up to 96% of life on earth drove jellyfish extinct (Gershwin 2013). I could not find a jellyfish cookbook. You could create the first!

Somebody Send the Bat Signal!

Healthy topsoil and freshwater are essential for our future national security, indispensable if we are to grow biomass for food, infrastructure, and energy. Yet the US exports 112,000 barrels per day of fuel ethanol causing erosion, water depletion, pollution, and eutrophication of waterways in our own country (EIA 2019).

If only Batman could save us from these polluting, erosion-prone, aquifer draining supervillain crops that take up half our cropland. Better yet, do not send Batman, send bats to eat crop pests and lower pesticide use. Please write your representatives to repeal the renewable fuel standard and drive a wooden stake through the vampire ethanol industry. We are just trading petroleum for an equal amount of alcohol, with no net effect except losing a huge amount of our most precious resource. Soil!

References

Al-Kaisi M (2000) Soil erosion: an agricultural production challenge. Integrated Crop Management. Iowa State University. https://crops.extension.iastate.edu/encyclopedia/soil-erosion-agricultural-production-challenge. Accessed 9 Nov 2020

DiBartolomeis M, Kegley S, Mineau P, et al (2019) An assessment of acute insecticide toxicity loading (AITL) of chemical pesticides used on agricultural land in the United States. PLOS ONE

EIA (2019) The united states exported a record volume of ethanol in 2018 for a second consecutive year. U.S. Energy Information Administration. https://www.eia.gov/todayinenergy/detail.php?id=39212. Accessed 9 Nov 2020

Fargione J, Hill J, Tilman D et al (2008) Land clearing and the biofuel carbon debt. Science 319:1235–1238

Fernandez-Cornejo J, Nehring R, Osteen C, et al (2014) Pesticide use in U.S. agriculture: 21 selected crops 1960–2008. United States Department of Agriculture

Fingerman K, Kammen D, O'Hare M (2008) Integrating water sustainability into the low carbon fuel standard. University of California, Berkeley, CA

Gerbens-Leenes W, Hoekstra AY, van der Meer TH (2009) The water footprint of bioenergy. Proc Natl Acad Sci 106:10219–10223

Gershwin L (2013) Stung! On jellyfish blooms and the future of the ocean. University of Chicago Press

Ippolito J, Al-Kaisi M (2019) The dirt on soil loss from the Midwest floods. The Conversation. https://theconversation.com/the-dirt-on-soil-loss-from-the-midwest-floods-114423. Accessed 9 Nov 2020

Lark T, Bougie M, Spawn S, et al (2018) Cropland expansion in the United States, 2008–2016. Gibbs Land Use and Environment Lab. http://www.gibbs-lab.com/wp-content/uploads/2018/11/Land-conversion-research-brief_11.2.18.pdf Accessed 9 Nov 2022

Lee L (2020) 54 million Americans are going hungry. Here's how you can make sure you eat. CNN. https://www.cnn.com/2020/10/24/us/how-to-get-food-assistance-hunger-pandemic-iyw-trnd/index.html. Accessed 9 Nov 2020

Levy MC, Neely WR, Borsa AA et al (2020) Fine-scale spatiotemporal variation in subsidence across California's San Joaquin Valley explained by groundwater demand. Environ Res Lett 15:104083

Mathews T (2014) Row crops are susceptible to soil erosion. Farm Horizons. http://www.herald-journal.com/farmhorizons/2014-farm/soil-erosion.html. Accessed 9 Nov 2020

Maupin MA, Kenny JF, Hutson SS, et al (2014) Estimated use of water in the United States in 2010. U.S. Geological Survey Circular 1405

Mclaughlin S, Walsh ME (1998) Evaluating environmental consequences of producing herbaceous crops for bioenergy. Biomass Bioenergy 14:317–324

Murphy DJ, Hall CAS, Powers B (2011) New perspectives on the energy return on investment of corn based ethanol. Environ Dev Sustain 13(1):179–202. http://netenergy.theoildrum.com/node/6760

NRC (2003) Frontiers in agricultural research: food, health, environment, and communities. National Research Council, National Academy of Sciences. https://doi.org/10.17226/10585

NRC (2014) The nexus of biofuels, climate change, and human health: Workshop summary. Institute of Medicine. National Research Council, National Academies Press

Padgitt M, Newton D, Renata P, et al (2000) Production practices for major crops in U.S. Agriculture, 1990–97. Resource Economics Division, Economic Research Service, U.S. Department of Agriculture

Patzek TW (2004) Thermodynamics of the corn-ethanol biofuel cycle. Crit Rev Plant Sci 23:519–567

Philpott T (2019) The hidden catastrophe of the Midwest's floods. Mother Jones. https://www. motherjones.com/environment/2019/03/the-hidden-catastrophe-of-the-midwests-floods/. Accessed 9 Nov 2020

Pimentel D (2003) Ethanol fuels: Energy balance, economics and environmental impacts are negative. Nat Resour Res 12:127–134

Pimentel D (2006) Soil erosion: a food and environmental threat. Environ Dev Sustain 8:119–137

Piñeiro G, Jobbágy EG, Baker J et al (2009) Set-asides can be better climate investment than corn ethanol. Ecol Appl 19:277–282

Powers SE (2005) Quantifying cradle-to-farm gate life-cycle impacts associated with fertilizer used for corn, soybean, and stover production. National Renewable Energy Laboratory, U.S. Department of Energy

RCN (2011) Heavy agricultural machinery can damage the soil, Nordic researchers find. Research Council of Norway, ScienceDaily. https://www.sciencedaily.com/releases/2011/05/110505083737.htm. Accessed 9 Nov 2020

Schulz WG (2007) The costs of biofuels. Chem Eng News 85:12–16

Searchinger T, Heimlich R, Houghton RA et al (2008) Use of U.S. croplands for biofuels increases greenhouse gases through emissions from land-use change. Science 319:1238–1240

Sullivan P (2004) Sustainable Soil management. Soil Systems Guide. ATTRA. https://soilandhealth.org/wp-content/uploads/01aglibrary/010117attrasoilmanual/010117attra.html. Accessed 9 Nov 2020

Tomson B (2007) For Ethanol, U.S. may boost corn acreage. Wall Street Journal. https://www.wsj.com/articles/SB117086485800701031. Accessed 9 Nov 2020

Troeh F, Thompson LM (2005) Soils and soil fertility, 6th edn. Wiley-Blackwell

Uri ND (2000) Global climate change and the effect of conservation practices in US agriculture. Environ Geol 40:41–52

USDA (2018) USDA reports soybean, corn acreage down. United States Department of Agriculture. Acres: Corn: 81.8 million soybeans 89.6 million

USDA (2019) Soybean production up in 2018, USDA reports. U.S. Department of Agriculture

USDA (2020) SNAP data tables. Latest Available Month April 2020 State Level Participation & Benefits by person. United States Department of Agriculture. https://www.fns.usda.gov/pd/supplemental-nutrition-assistance-program-snap. Accessed 10 Nov 2020

Zattara EE, Aizen MA (2019) Worldwide occurrence records reflect a global decline in bee species richness. Cold Spring Harbor Lab. https://doi.org/10.1101/869784

Chapter 24
Corn Ethanol. Why?

I have explained why I think we will return to wood and biomass for thermal heat and infrastructure—for steel, cement, manufacturing, transportation, cooking, and heat. We will rely on biomass, just like back in the day, civilization such as it was before fossil fuels. Why? Because it may be the only thing able to do the job, or some of it, that is left. I hope I am wrong, but if I am right, then we need to understand what role liquid biofuels might play in the future. At this juncture, it is time to look at terrestrial and aquatic plants, what fuels they can make, and if they will be useful in Wood World.

What Is the Point of Making Ethanol?

Surely there must be a yet to be revealed good reason for the federal government to mandate that transportation fuels include 11.56% biofuels by volume (the 2020 requirement). We are willing to tolerate the damage from corn and soy biofuels, the water depletion, erosion, loss of biodiversity, habitat loss, toxic algal blooms, pollution of waterways, greenhouse gas emissions, road and bridge damage, the loss of valuable petroleum resources, and increased cost for food and feed (EPA 2018). So, there must be some redeeming value for this biofuel mandate. Right? Wrong.

But do not we reduce the need for foreign oil? Not if the energy return is low or negative, and certainly not if the energy to fix the harm was included in the calculation.

The biofuels mandate does not lend a push to the trucking industry. While Rudolf Diesel originally thought his diesel engine would run on biological oils, and Henry Ford spent much of his life trying to invent soybean fuels (Shurtleff and Aoyagi 2011), to this day heavy-duty trucks can not run on ethanol or diesohol (Beer et al. 2001). Diesel engines just can not stomach ethanol (Bika et al. 2009).

A. J. Friedemann, *Life after Fossil Fuels*, Lecture Notes in Energy 81, https://doi.org/10.1007/978-3-030-70335-6_24

One argument for the use of ethanol: It is used to oxygenate fuels. Oxygenates can enhance fuel combustion and thereby reduce exhaust emissions. MTBE once performed this role. But MTBE is a carcinogen that easily mixes with water and contaminated aquifers, so MTBE was banned. There are ways to oxygenate fuels at the refinery without ethanol, although in some cases ethanol can be the preferred option for boosting octane (Rapier 2018).

Not all cars need these oxygenated fuels. In fact, automobiles built since 1995 do not need them. Thanks to advances in combustion technology and emissions controls (SC 2020), they burn clean without that assistance.

Better miles per gallon? No, the opposite. Ethanol has 30% less energy per unit of volume than gasoline. So, if the idea was to stretch out gasoline supplies by dilution with ethanol, it does not work. We just refill our gas tanks more often.

Greenhouse Gas Emissions?

Remember, in the beginning, ethanol was sold as being carbon neutral and good for the climate. The concept was that corn and soybean took carbon dioxide out of the air when grown, returned it to the atmosphere when used as a fuel, and so were carbon neutral. That did not account for what it would take to put Midwest corn and soy in your gas tank. That complex agricultural, refining and transportation process unleashes a lot of additional carbon, nitrous oxide, and other greenhouse and toxic gases into the atmosphere.

But does not ethanol mitigate climate change? No, ethanol is worse than gasoline. For every billion ethanol-equivalent gallons of fuel produced and combusted in the US, an equivalent amount of petroleum was used to produce it (see Chap. 22). Hill et al. (2009) estimated that the combined climate change and health costs are $469 million for gasoline, and $472–$952 million for corn ethanol (Hill et al. 2009).

Yes, climate was supposed to be the why. The Renewable Fuel Standard (RFS) of 2007 mandated that only biofuels which lowered greenhouse gas emissions 20% or more than petroleum emissions were eligible to be added to gasoline or diesel.

Surprise, a giant loophole existed. Ethanol plants in operation or under construction by mid-December 2007 did not need to comply. These grandfathered plants, exempt from emissions reduction requirements, have produced the majority of ethanol to date (GAO 2019).

When all emissions are added up, ethanol releases up to twice as much greenhouse gas per unit of energy delivered as gasoline (Hertel et al. 2010; Mullins et al. 2011; NRC 2011; Searchinger et al. 2008; Plevin et al. 2010; NRC 2014; Liska et al. 2014). Find that hard to believe? Can you find research that says the opposite? Yes, you can. That is because what gets included in the growing, refining, and distribution of ethanol is not consistent between studies. Let us take a careful look at this.

The idea that biofuels generate less CO_2 than gasoline stems from the fact that biofuels are derived from plants that absorb carbon dioxide. But land typically supports plant growth regardless of whether it is being used to grow corn or not. Corn grown for ethanol for use in gasoline has a net benefit of storing around 3 tons of carbon dioxide per hectare. But if the land had not been used for ethanol, we would be better off. If reforested, then 7.5–12 tons of CO_2 would be stored per hectare. A corn ethanol field, formerly a forest, will emit 12–35 tons of CO_2 per hectare a year for 30 years (NRC 2014). By contrast, a wetland stores 81–216 tons of carbon per acre (TCF 2020). In sum, corn does not sequester carbon but recycles it at best, releasing CO_2 when made into ethanol, and absorbing CO_2 in the next corn crop.

Every year when land is tilled or cleared to grow crops, greenhouse gases are emitted from the soil. A carbon storehouse, soil stores 4.5 times more carbon than vegetation (Lal 2004).

Agriculture emits 30% of all global greenhouse gas emissions. Even if we stopped using fossil fuels today, the food system alone would exceed the 1.5 °C (2.7 °F) emissions limit between 2051 and 2063 and 2 °C by 2100. That is, unless we reduce food waste and loss by 50%, use less fertilizer, increase production efficiency, and other measures (Clark et al. 2020). Growing more biomass for fuel and food will only make emissions greater, especially if forests and other land is cleared to do so.

And here is the real kicker. Agriculture emits 80% of all nitrous oxide (N_2O) with a global warming potential 300 times higher than CO_2 and destroys stratospheric ozone while it is at it (Melillo et al. 2009; Mosnier et al. 2013). The use of fertilizer triggers these emissions. N_2O emissions have increased substantially due to greater nitrogen fertilizer use, especially for corn, which uses greater amounts than most other crops.

In addition to CO_2 and N_2O, corn ethanol as part of your car's fuel mix emits other air pollutants and in greater quantity than does gasoline. These include carbon monoxide [CO], sulfur dioxide [SO_2], particulate matter [PM], ozone [O_3], hazardous volatile organic compounds, and other pollutants, such as benzene and formaldehyde. Pollutants can damage human health, causing cancer, cardiovascular disease, respiratory irritation, and birth defects, and harm the environment in reduced visibility, acidification of water and soils, and damage to crops (NRC 2011).

E15 Increases the Damage by 50%

In 2011, E15 was allowed (15% ethanol 85% gasoline), opening the door for 50% more pasture, range, and forests to be converted to corn. The 2011, E15 allowance excluded its use in the summer when it can increase smog. The near decade-long ban on summertime E15 use has ended. Donald Trump, desperate for Midwest votes after the trade war he started hurt farmers and the ethanol industry, took the summer restriction away. The ethanol industry used to sell 20% of its ethanol to China. Because of the trade war, China jacked up tariffs on ethanol 30% in 2017 and another 40% in 2018 (Reuters 2019).

E15 can harm motorcycles, school buses, delivery trucks, boats, snowmobiles, chain saws, lawnmowers, and cars built before 2001. And the planet!

Transportation of Ethanol from Midwest to Coasts a Waste of Diesel Energy

The vast majority of ethanol is produced in the Midwest, yet 80% of the US population lives within 200 miles of the coast. It is expensive to move it from the Midwest since ethanol can not travel in pipelines, which costs up to 20 times less than trucks and 5 times less than rail. Ethanol is too corrosive to travel in pipelines (Curley 2008). So instead, low EROI ethanol is moved with high EROI petroleum diesel trucks, locomotives, and barges. Remember EROI: Energy return on energy invested. Think of it as a measure of the smart use of energy.

In 2017, 15.8 billion gallons of ethanol were transported. About 65% was moved by 340,000 railcars carrying 30,000-gallons, and 690,000 trucks with 8000-gallon tanks.

This, in turn, damaged thousands of miles of rural roads. A truckload of ethanol weighs 77,720 pounds, close to the maximum road weight of 80,000 pounds. Just one heavy truck can do more damage than nearly 10,000 cars (GAO 1979).

State and local roads were not built for so many heavy trucks. It is not just ethanol trucks damaging roads, it is also trucks hauling food and feed corn to grain elevators and DDGS byproduct from ethanol plants.

Ethanol Raises Food Prices and Harms People and Businesses

The Renewable Fuel Standard (RFS) mandating ethanol has led to a shortage of corn for food and animal feed. From 2007 to 2012, prices were driven up so much that farmers planted 17 million new acres of corn rather than soybeans, wheat, hay, cotton, and other crops, driving their prices up to all-time records as well. Cattle feed prices were so high that herds were culled to levels not seen in 60 years, causing beef prices to rise an incredible 60% from 2007 to 2012 (Landstreet 2018).

Restaurants were also affected because corn, meat, and other crops rose in price. It appears the interests of Archer Daniels Midland (ADM), Cargill, and 3.2 million farmers were favored over those of us who eat food. That includes the 15.6 million Americans who work in the restaurant industry—about one in ten US workers (PWC 2012; NRA 2020).

Ethanol Was Mandated to Enrich the Wealthiest Companies and Farmers in the Midwest

Spoiler alert: Since so much about the required use of ethanol does not make any sense, it appears the main purpose was to enrich wealthy corporations and corporate farms in the Midwest. The 2007 RFS mandate has steered tens of billions of taxpayer dollars to the corn ethanol industry, especially Archer Daniels Midland, Cargill, and POET LLC (TCS 2014). The RFS is here to stay, despite the harm, because of the US electoral college. The six states that make 75% of ethanol comprise 10% of the electoral votes. Iowa makes 25% of ethanol and is the first-in-the-nation presidential caucus (EB 2018).

Ethanol refineries would probably go out of business without the RFS government mandate that forces ethanol into gasoline. The RFS includes not only a mandate but a treasure chest of rewards including tax credits, state and federal subsidies, entitlement programs, the renewable identification number program, crop and ethanol blender insurance, Department of Energy clean cities program, tariff protections, road damage paid by taxpayers, and dozens of other subsidies (TCS 2015). From 1980 to 2011, they add up to about $57 billion (Stevens et al. 2016). On top of that, local governments give away tax benefits to lure ethanol refineries to their towns.

The game may be fixed but there are always jokers in the deck. Despite all these gifts, farmers and ethanol producers do not always come out ahead. Ethanol refineries need low corn prices to make a profit. Following corn price spikes in 2008 and 2012, many corn ethanol plants went offline, unable to run profitably (Salter 2013). Meanwhile, the rising costs of inputs like fertilizer drove profits down (Stevens et al. 2016).

Conclusion

The nineteenth-century western farmers converted their corn into whiskey to make a profit (Rorabaugh 1979). Archer Daniels Midland (ADM), a large grain processor, came up with the same scheme in the twentieth century. But ethanol was a product in search of a market, so ADM spent three decades relentlessly lobbying for ethanol to be used in gasoline (Barrionuevo 2006). Today ADM and other corporations continue to profit from ethanol and government subsidies. We are being force-fed ethanol!

The US Department of Agriculture has known for a long time that corn bankrupts soil. A 1911 report reads as follows: "When the rich, black, prairie corn lands of the Central West were first broken up, it was believed that these were inexhaustible lands. So, crop after crop of corn was planted on the same fields. There came a time after 15 or 20 years, when the crop did not respond to cultivation; the yields fell off and the lands that once produced 60–70 bushels per acre annually dropped to 25–30 bushels. With passing years, the soil became more compact, droughts more injurious, and the soil baked harder and was more difficult to handle. Continuous corn culture has no place in progressive farming. It is a shortsighted policy and is suicidal on lands that have been long under cultivation." (Smith 1911).

Some things do not change. We never learn.

References

Barrionuevo A (2006) A bet on ethanol, with a convert at the helm. New York Times. https://www. nytimes.com/2006/10/08/business/yourmoney/08adm.html. Accessed 6 Nov 2020

Beer T, Grant T, Morgan G, et al (2001) Comparison of transport fuels final report. Australian Greenhouse

Bika AS, Franklin L, Olson AL, et al (2009) Ethanol utilization in a diesel engine. University of Minnesota, Department of Mechanical Engineering

Clark MA, Domingo NGG, Colgan K et al (2020) Global food system emissions could preclude achieving the 1.5° and 2°C climate change targets. Science 370:705–708

Curley M (2008) Can ethanol be transported in a multi-product pipeline? Pipeline Gas J 235:34

EB (2018) Doubling down on the biofuel boondoggle. Editorial Board, Washington Post.   https://www.washingtonpost.com/opinions/doubling-down-on-the-biofuel-boondoggle/2018/10/14/3092cd7e-ccbc-11e8-920f-dd52e1ae4570_story.html.   Accessed   6 Nov 2020

EPA (2018) Biofuels and the Environment: The Second Triennial Report to Congress. U.S. Environmental Protection Agency, Washington, DC, EPA/600/R-18/195

GAO (1979) Excessive truck weight: an expensive burden we can no longer support. U.S. Government Accountability Office

GAO (2019) Renewable fuel standard. Information on likely program effects on gasoline prices and greenhouse gas emissions. U.S. Government Accountability Office

Hertel TW, Golub AA, Jones AD et al (2010) Effects of US maize ethanol on global land use and greenhouse gas emissions: Estimating market-mediated responses. Bioscience 60:223–231

Hill J, Polasky S, Nelson E et al (2009) Climate change and health costs of air emissions from biofuels and gasoline. U.S. Proc Natl Acad Sci USA 106:2077–2082

Lal R (2004) Soil carbon sequestration impacts on global climate change and food security. Science 304:1623–1627

Landstreet T (2018) Biofuel mandates are a bad idea whose time may be up. Wall Street J

Liska AJ, Yang H, Milner M et al (2014) Biofuels from crop residue can reduce soil carbon and increase CO2 emissions. Nat Clim Chang 4:398–401

Melillo JM, Reilly JM, Kicklighter DW et al (2009) Indirect emissions from biofuels: how important? Science 326:1397–1399

Mosnier A, Havlík H, Valin H et al (2013) The net global effects of alternative U.S. biofuel mandates: fossil fuel displacement, indirect land use change, and the role of agricultural productivity growth. Energy Policy 57:602–614

Mullins KA, Griffin WM, Matthews HS (2011) Policy implications of uncertainty in modeled life-cycle greenhouse gas emissions of biofuels. Environ Sci Technol 45:132–138

NRA (2020) Restaurant industry factbook. National Restaurant Association. https://restaurant.org/ downloads/pdfs/research/soi/2020-state-of-the-industry-factbook.pdf. Accessed 10 Nov 2020

NRC (2011) Renewable fuel standard: potential economic and environmental effects of U.S. biofuel policy. National Resource Council, National Academies Press

NRC (2014) The nexus of biofuels, climate change, and human health: Workshop summary. Institute of Medicine. National Resource council, National Academies Press

Plevin RJ, O'Hare M, Jones AD et al (2010) Greenhouse gas emissions from biofuels' indirect land use change are uncertain but may be much greater than previously estimated. Environ Sci Technol 44:8015–8021

PWC (2012) Federal ethanol policies and chain restaurant food costs. PricewaterhouseCoopers. https://smarterfuelfuture.org/assets/content/resources/Federal_Ethanol_Policies_and_Chain_ Restaurant_Food_Costs.pdf. Accessed 10 Nov 2020

Rapier R (2018) Addressing misconceptions from Senator Grassley's ethanol editorial. Forbes. https://www.forbes.com/sites/rrapier/2018/05/13/addressing-misconceptions-from-senator-grassleys-ethanol-editorial/?sh=3078995315ee. Accessed 10 Nov 2020

Reuters (2019) Trade war hurting the U.S. ethanol industry 'badly': U.S. Grains Council. https://www.reuters.com/article/us-usa-trade-china-ethanol/trade-war-hurting-the-u-s-ethanol-industry-badly-u-s-grains-council-idUSKCN1SL2P1. Accessed 10 Nov 2020

Rorabaugh WJ (1979) The alcoholic republic: an American tradition. Oxford University Press

Salter J (2013) Corn shortage idles 20 ethanol plants nationwide. https://www.usatoday.com/story/news/nation/2013/02/10/corn-shortage-idles-plants-nationwide/1906831/. Accessed 10 Nov 2020

SC (2020) Gasoline additive MTBE. Sierra Club. https://www.sierraclub.org/policy/energy/gasoline-additive-mtbe. Accessed 10 Nov 2020

Searchinger T, Heimlich R, Houghton RA et al (2008) Use of U.S. croplands for biofuels increases greenhouse gases through emissions from land-use change. Science 319:1238–1240

Shurtleff W, Aoyagi A (2011) Henry Ford and his researchers. History of their work with soybeans, soyfoods, and chemurgy (1928–2011). Soyinfo Center. http://www.soyinfocenter.com/pdf/145/Ford.PDF Accessed 12 Dec 2020

Smith CB (1911) Rotations in the corn belt. Yearbook of the Department of Agriculture, pp 334–336

Stevens L, Simmons RT, Yonk RM (2016) Ethanol and renewable fuel standard. Utah State University. https://www.strata.org/rfs/. Accessed 10 Nov 2020

TCF (2020) Green infrastructure and ecosystem services assessment. Houston-Galveston, The Conservation Fund. http://www.conservationfund.org/images/projects/files/Houston_Galveston_Report.pdf. Accessed 10 Nov 2020

TCS (2014) Political footprint of the corn ethanol lobby. Taxpayers for Common Sense. https://www.taxpayer.net/agriculture/updated-political-footprint-of-the-corn-ethanol-lobby/. Accessed 10 Nov 2020

TCS (2015) Federal subsidies for corn ethanol and other corn-based biofuels. Taxpayers for Common Sense. https://www.taxpayer.net/energy-natural-resources/federal-subsidies-for-corn-ethanol-and-other-corn-based-biofuels/. Accessed 10 Nov 2020

Chapter 25
Biodiesel from Algae

The green technology sector once fell in love with algae. We do not know if the attraction was mutual. But we do know that the plan was that the two would get together and produce a third-generation biofuel. Since algae can produce many times more biomass per square foot than terrestrial plants, algal biofuels hold a great deal of promise. Importantly, they are best suited for making biodiesel, the essential fuel. Ships, trucks, and trains are the backbone of civilization, and they depend on diesel.

Think of it, a world powered by pond scum!

We know how to grow algae, though there are no successful commercial fuel production facilities. The vast majority of commercial algal products are used for nutritional supplements, cosmetics, and other products.

The main reason algal fuels are not being produced is the problem of "pond crash."

In practice, about a third of the time all of a pond's algae die within 3 months (Park et al. 2011). It does not take much head-scratching to figure out why: The pond is wide open to invading algae predators via wind, rain, snow, insects, waterfowl, and animals. Among the predators are zooplanktons. Each one can eat 200 algae a minute and crash a pond in less than 2 days (SNL 2017).

They are not the only marauders. There are also killer viruses, fungi, diseases, and amoebas. And open ponds are ideal breeding territory for mosquitoes, which prey not only on us but also on algae.

Algae are Goldilocks creatures. Everything must be just right. They can easily be killed or have their growth stunted by too much heat, cold, evaporation, pH level, saline level, UV, lack of nutrients, or too much of a nutrient (DOE 2010).

Nor will just any algae do. For algal biofuels, the goal is to use obese algae with at least 60% fat to make as much biodiesel as possible for the least cost. But usually,

This chapter, updated here, was originally published by Gavin Publishers. Friedemann, A. J. 2019. Archives of Petroleum & Environmental Biotechnology Volume 4; Issue 2. https://doi. org/10.29011/2574-7614.100055

A. J. Friedemann, *Life after Fossil Fuels*, Lecture Notes in Energy 81, https://doi.org/10.1007/978-3-030-70335-6_25

tougher, leaner, faster reproducing algae get into the pond and outcompete the plump ones (Hall and Benemann 2011).

If only a microscopic border patrol could keep them out. Why not build walls around the ponds! Oh wait, there are walls. Screens have not worked, nor pesticides, since microscopic predators develop immunity quickly because they reproduce in just a day or two.

If algal biofuels are the future, then grains and oilseeds are the current biofuel feedstock. It is clear that terrestrial biomass does not scale up enough to run the world on biofuels. In Europe, it has been estimated that two billion metric tons of grains and oilseeds are grown a year, but that 15 billion metric tons, 7.5 times as much, would be needed to replace oil with biofuels (Richard 2010).

Currently, only 15,000 tons a year of algae are produced (DOE 2016). Compared to current biofuels, algae are tremendously expensive to produce, ranging from $719 to $3000 per dry ton, versus switchgrass, corn stover, and other land biomass costing $30–$60 per dry ton (DOE 2016).

Where Is the Flat 1200-Acre Land for Ponds?

The ponds for growing algae have to be huge, about 1200 acres of very flat land (less than a 1% grade) containing multiple 10-acre or more ponds for economies of scale, ideally near a city to reduce the cost of delivery. Ideally, this land also has impermeable soil below to reduce the energy required to line and seal the ponds and prevent seepage of toxins into the groundwater.

Ponds also need to be large because they can not be deep since sunlight does not penetrate algal growths for more than a couple of inches (Wald 2012). Too much sun is also harmful since algae can suffer oxidative damage. Many species protect themselves by inhibiting photosynthesis, which constrains growth (DOE 2010).

Algae feed on CO_2, which must be pumped into the ponds. CO_2 is a huge cost for algae farming, up to 25% of overall costs. Yet there are very few industries emitting excess CO_2 that also have 1200 acres of very flat land nearby, nor wastewater plants to provide water for that matter (DOE 2016).

The fairy-tale princess may have been overly sensitive to a pea under her mattress, but algae are even more delicate. A proper berth for them would need at least 2800 h of sunshine per year since sunlight is the most important factor in algal growth. That much sun exists in only eight states. There would ideally be 40 inches of rain and low evaporation rates, not likely in Arizona and the other suitable arid states. In addition, the ponds do best when the average temperature is 55 F or more, at least 200 days are above freezing, there is little wind so that predators, dust, and sand are not swept into ponds, and heavy rain, flooding, hail, tornadoes, or hurricanes are rare. Such fussbudgets!

There is competition for the use of flatlands. Algal ponds compete with agriculture and recreation, as well as solar facilities, which can produce far more energy than algae over their lifespan on considerably less land (Wigmosta et al. 2011; NRC 2012b).

Where Is the Water?

Large-scale algal biofuel production is likely to require as much water nationally as large-scale agriculture (DOE 2010). Wigmosta et al. (2011) estimated that to produce 220 billion liters of algal biofuels—that would equate to 28% of US transportation fuel—the evaporative loss from ponds would be 312 trillion liters per year. That is about twice the quantity of water used for irrigated agriculture in the US (Dieter et al. 2018).

An advantage of algae over land plants is that the water can be saline, brackish, wastewater, and low quality. The problem is that the water being evaporated is fresh, and continuing to use low-quality water to refresh the pond can introduce and concentrate killer microbes, heavy metals, chemicals, salts, toxins, and other harmful materials (DOE 2010). This would also render any coproducts from algal sludge unsuitable for animal feed. If wastewater is to be used, there are not many wastewater treatment plants with thousands of acres of cheap flat land nearby where ponds could be created.

Carbon Dioxide Problems Coming and Going

Unlike plants, which can make use of CO_2 in the air, commercial algae production requires concentrated CO_2 because not enough CO_2 from the air penetrates the water (DOE 2010; Williams and Laurens 2010; NRC 2012a). Coal-fired power plants would seem to be an ideal source for this CO_2. Algae however can only use CO_2 when the sun is shining, and not at night. Thus, in terms of the hope of using algal ponds to limit greenhouse gas emissions from coal power plants and other CO_2 emitting industries, algal ponds would not be able to offset more than 20–30% of the total power plant emissions (Brune et al. 2009).

There is another CO_2 issue with algal ponds. Ninety percent of the CO_2 pumped into an algal pond will bubble up to the surface and into the air, resulting in substantially higher net emissions from algal biofuels than petroleum, according to several studies (DOE 2010; Wald 2012; NRC 2012b). The 2007 renewable fuel standard mandated that only biofuels which lowered greenhouse gas emissions 20% or more beyond petroleum emissions were qualified to be added to gasoline or diesel.

Microscopic Algae Are as Voracious as Food Crops

The amount of nutrients required to grow enough algae to produce just 5% of transportation fuel could be as high as required by large-scale agriculture (DOE 2010; NRC 2012b). To produce just 5% of the transportation fuels used in the United States, an alga with an oil content of 20% would need more nitrogen than the US consumes today on crops. This same quantity of algal biofuels also would require phosphorus equivalent to up to half of what is currently consumed by US agriculture (NRC 2012a). There is a danger of phosphorous depletion as soon as 2080–2100 (Vaccari 2009; Smil 2000).

Recycling algae to get the nitrogen and phosphorus back is not easy. It is expensive and energy-intensive to separate them from the dead algae after their oil has been removed to make biodiesel, and 20–40% cannot be recovered.

Where Is the Energy?

The main reason to make algal biodiesel is to provide a substitute for petroleum diesel. Other metrics such as CO_2 sequestration, byproducts, and GHG emissions are not as relevant. All that matters is that the EROI (energy return on investment) may need to be as high as 10 or more to maintain our current level of civilization (Murphy et al. 2011; Lambert et al. 2014).

An absolute showstopper is the very negative EROI of algal biofuels: Far more fossil fuel energy is needed to build and grow the algae than the energy contained in the algal fuel. The energy for water management alone is seven times more than the algal biodiesel created, and water management is just a fraction of the overall energy inputs (Murphy and Allen 2011).

As with corn ethanol, estimates of EROI for algal biofuels vary wildly, ranging from negative to positive (NRC 2012b). As with ethanol, proponents who find positive results rely on adding the energy of the byproduct, in this case algal sludge. The NRC (2012b) reports that one proponent, Sander and Murthy (2010) gave an "energy credit for using algae residuals 10 times larger than the energy content of the produced biodiesel." Yet even then the EROI was a trivial 1.77–3.33. Other studies found that it takes three to eight times as much fossil fuel energy inputs as the energy contained in the algal biofuel. Closed bioreactors can use up to 57 times more fossil fuel energy (NRC 2012b).

Sorry to Let the Air Out of Your Balloon

There may be those among you who have drunken the algal Kool-Aid. I have tried to sober you up. If there are any incorrigible algal optimists left reading this, consider a subset of the steps and inputs needed to make algal biofuel. I have summarized the process below and Capitalized Each Action or Object that requires fossil fuel energy.

Algae need light to survive and grow. To get adequate light, the pond can only be a few inches deep, so ponds have to be large, which adds to Construction and land costs. Water needs to be Pumped into and between ponds. The algae at the top hog most of the sunlight, so the water must be constantly Stirred, Pumped, and Circulated. On a hot day, an inch or more water evaporates, so more water must be Pumped in. After a pond crash the pond must be thoroughly Cleaned. CO_2 must be Collected, Compressed, and Pumped into Pipelines to deliver CO_2 to the facility via Tubes at the bottom of ponds, which can get clogged, requiring regular Cleaning. · Agitators, Aerators, and Fountains must run constantly to distribute nutrients and CO_2 and to discourage mosquitos from breeding.

An algal biofuel facility is made of Cement, Plastic, Pumps, Centrifuges, Chemicals, Filters, Pond Liners, CO_2 Waste Treatment Facilities, Drying Areas, Fuel Processing, Transport, and Storage Infrastructure. Nitrogen, Phosphorus, and other nutrients must be Produced, Transported, and Distributed in the ponds. Treating the wastewater requires Decontamination, Disinfection, and Removal of heavy metals. Water must be Heated or Cooled to maintain an optimal temperature. It also takes energy to Monitor and Keep pH levels, saline levels, nutrient, and water levels at optimal levels.

To make the algal fuel, algae are Pumped through each of these steps: Harvest, Filter, Sieved, Dry, and Extract oil. Recycle nutrients, Dispose of wastewater. Getting the water out is a huge part of the energy used: Algae are single cells suspended in water at concentrations below 1% solids, whereas land plants are often over 40% solids. The energy to Concentrate and Dry the algae commercially is far greater than the energy contained in the algae (DOE 2010). Extract the oil in the algae. Transform this oil into biodiesel (many steps not listed here). Finally, Store, Transport, Blend, Deliver, and Dispense algal biodiesel.

Protect Algae from Crashes by Sheltering Them in Photobioreactors

You might think algae could be protected from predators in the long glass or plastic tubes of a photobioreactor, but microscopic creatures can also get into them and form bacterial biofilms that slow down water flow and reduce the light. However much trouble ponds may be, photobioreactors are far more problematic and expensive, have never been scaled up to a commercial level, cost more, and use far more energy than ponds. They can not be sterilized and must be cleaned, they need energy-intensive temperature, pH, dissolved oxygen, and CO_2 controls. They are far from being commercial. Bottom line: they require far more energy than open ponds and studies have found all of them to have a negative energy return on invested (DOE 2010; NRC 2012b).

Conclusion: Algae May Be Green, but They Are Not Clean

Discharging untreated water from an algal pond can lead to eutrophication of water-ways, contaminate groundwater, salinize freshwater, harm wildlife, and be a source of heavy metals, herbicides, algal toxins, and industrial effluents. Untreated water may escape in a flood, earthquake, tornado, high rainfall, and when the pond leaks or breaks. If a foreign or bio-engineered algal species escapes, it could threaten local and regional ecosystems by displacing native species and causing dense algal blooms that block sunlight.

Algae also compete with agriculture for very flat land.

There are simply too many showstoppers. Algae are greedy little bastards, need-ing more water, nitrogen, and phosphorus than corn or soybeans, placing unsustain-able demands on energy, water, and nutrients (NRC 2012b).

Clearly, algal fuels are far from being commercial, unless you can get the mili-tary to pay for them that is. In 2009, the Pentagon spent $424 a gallon on algae oil (Cardwell 2012).

Scientists, entrepreneurs, and the US government have been trying to make algal biofuels for over 45 years, ever since the 1970 oil shocks, and have studied over 3000 kinds of algae for their biofuel potential. It has been a long romance. But after decades of research, the Department of Energy gave up and stopped funding in 1995 (Sheehan et al. 1998).

Do not be misled by related recent Department of Energy research. It is focused on cleaning up CO_2 from power plants to lower greenhouse emissions (DOE 2016), not producing biofuels to keep trucks running. Those trucks and that diesel make civilization as we know it possible.

References

Brune DE, Lundquist TJ, Benemann JR (2009) Microalgal biomass for greenhouse gas reductions: potential for replacement of fossil-fuels and animal feeds. J Environ Eng 135:1136–1144
Cardwell D (2012) Military spend on biofuels draws fire. New York Times
Dieter CA, Maupin MA, Caldwell RR, et al (2018) Estimated use of water in the United States in 2015. U.S. Geological Survey Circular 1441. https://doi.org/10.3133/cir1441
DOE (2010) National algal biofuels technology roadmap. U.S. Department of Energy, Washington, DC

DOE (2016) 2016 Billion-ton report. Advancing domestic resources for a thriving bio economy. U.S. Department of energy

Hall CA, Benemann JR (2011) Oil from algae? Bioscience 61:741–742

Lambert JG, Hall CAS, Balogh S et al (2014) Energy, EROI and quality of life. Energy Policy 64:153–167

Murphy CF, Allen DT (2011) Energy-water nexus for mass cultivation of algae. Environ Sci Technol 45:5861–5868

Murphy DJ, Hall CAS, Dale M et al (2011) Order from chaos: a preliminary protocol for determining the EROI of fuels. Sustainability 10:1888–1907

NRC (2012a) America's energy future: technology and transformation 2009. National Research Council, National Academies Press

NRC (2012b) Sustainable development of algal biofuels. National Research Council, National Academies Press, Washington, DC

Park JBK, Craggs RJ, Shilton AN (2011) Wastewater treatment high rate algal ponds for biofuel production. Bioresour Technol 102:35–42

Richard T (2010) Challenges in scaling up biofuels infrastructure. Science 329:793–796

Sander K, Murthy GS (2010) Life cycle analysis of algae biodiesel. Int J Life Cycle Assess 15:704–714

Sheehan J, Dunahay T, Benemann J, et al (1998) A look back at the U.S. Department of Energy's aquatic species program: biodiesel from algae. U.S. Department of Energy, National Renewable Energy Laboratory

Smil V (2000) Phosphorus in the environment: natural flows and human interferences. Annu Rev Energy Environ 25:53–88

SNL (2017) Multilab project seeks toughest algae strains for biofuel. Sandia National Laboratories Biomass magazine. http://biomassmagazine.com/articles/14237/multilab-project-seeks-toughest-algae-strains-for-biofuel. Accessed 10 Nov 2020

Vaccari DA (2009) Phosphorus: a looming crisis. Sci Am 300:54–59

Wald ML (2012) Another path to biofuels. New York Times. https://green.blogs.nytimes.com/2012/11/23/another-path-to-biofuels-two-actually/. Accessed 10 Nov 2020

Wigmosta MS, Coleman AM, Skaggs RJ et al (2011) National microalgae biofuel production potential and resource demand. Water Resour Res 47(3)

Williams PJ, Laurens LM (2010) Microalgae as biodiesel and biomass feedstocks: Review and analysis of the biochemistry, energetics and economics. Energy and Environmental Science 3:554–590.

Chapter 26
Fill 'er Up with Seaweed

Seaweed grows … well, like a weed. And you can make a lot of things with it. Looking for a fresh vegetable? Have you tried bladderwrack? Want a fresh look? You will not recognize yourself after you give a try to "Restoring Marine Algae Overnight Face Mask."

Seaweed is used in many products, but not biofuels.

Seaweed, kelp, and macroalgae are the same thing—multicellular algae, cousins of the single-celled phytoplankton. There are over 10,000 species of brown, green, and red seaweed in marine environments, all of them edible. The most harvested seaweed is destined to become food, the rest will become fertilizer, supplements, ice cream, animal feed, cosmetics, carrageenan, agar, and other industrial products. But not biofuels.

Too Low-Fat for Biodiesel

Single-celled algae can be 50% lipids, but seaweed is skinny with a fat content of just 0.2–4%, so biodiesel can not be made. In that 45–75% of seaweed consists of carbohydrates, ethanol is possible.

Corn plants average 8 ft tall, growing about 1–1.5 inches a day. Giant brown kelp grows at a rate that dwarfs that of corn. This variety of kelp will give Jack and his beanstalk a run for their money, growing so fast it can reach 150 ft tall at a rate of 1–2 ft a day. This is due to a photosynthetic efficiency 12–16 times greater than corn. And they do it without fertilizers or pesticides.

Land plants stand tall because they have a strong cellulose structure like the steel framework of buildings. Seaweed has far less cellulose, which makes for an easier ethanol recipe which can use all of the plant, not just the grain like corn kernels. Simply pulverize and ferment.

A. J. Friedemann, *Life after Fossil Fuels*, Lecture Notes in Energy 81, https://doi.org/10.1007/978-3-030-70335-6_26

Because of their huge size, fast growth rate, and ease of mechanical harvest, giant brown kelp would be the ideal seaweed to provide large amounts of biomass for biofuel.

But scaling up kelp production for ethanol will not be easy. There is not enough kelp to begin with, since it is very fussy, needing cool water between 41 F and 57 F year-round, which is why you do not find kelp in the tropics. All species of kelp decline in water above 68 F where diseases such as black rot can attack and there are fewer nutrients for growth.

Where kelp can grow is further restricted to shallow coastal water where sunlight can penetrate, in a current with lots of nutrients, and a rocky bottom with which to attach. If detached, seaweed starves to death.

Just as floods, pests, and droughts destroy crops, seaweed too has its scourges. Kelp beds can be decimated by large storm waves that uproot them, harmed by disease, devoured by herbivores, and harmed by lack of sunlight from pollution, urban runoff, and siltation. Kelp can also be outcompeted by invasive species, harmed by dredging, trawling, boat traffic, or lack of nutrients.

In the future, climate change is likely to reduce kelp forests further due to greater storm frequency and severity, acidification, and warmer water. It is already happening in California's famous offshore kelp beds. In 2013, a mysterious disease devastated the starfish population. Starfish eat the sea urchins that eat kelp. A sea urchin population explosion occurred and the kelp forests began to disappear. Then came a heatwave, warming the water, which held so much less nitrogen the kelp could not grow fast enough to reach the surface for photosynthesis. Over 90% of the kelp along 200 miles of California's coast is now gone (Rogers-Bennett and Catton 2019).

Even in a good year only a fraction of kelp can be harvested since they are a keystone species that fosters fisheries and thousands of species that feed on or shelter in them. These ecosystems provide billions of dollars in terms of fishing, tourism, and ecosystem services such as waste treatment and reducing eutrophication and greenhouse gases. Kelp often grows in areas that can not be harvested, such as marine sanctuaries, near cities, recreation areas, or military sites.

Seaweed ethanol is not a happening thing. There is little if any seaweed ethanol being made because it is more profitable to sell seaweed for food and other products. That will be even more true as the population grows. Many nations see more seaweed production as a way to grow food supplies.

There are other problems with using seaweed for ethanol. The biggest is that seaweed is 70–90% water, whereas corn grain is a mere 15%. Getting the water out is energy-intensive, but all of it must be removed from ethanol before mixing it with gasoline. Water in the gasoline in your car gets you towed, your gas tank removed and flushed, and sets you back about $900 (Chatman 2018).

Growing seaweed artificially takes so much fossil energy to produce that several studies have found the EROI to be negative (Milledge and Harvey 2016).

On a commercial farm, seaweed cultivation needs acres of prime real estate—flat, low land near the ocean already served by roads, with room for tanks, a lab, an office, and other facilities. Energy is used to sterilize all seawater, power air blowers to add oxygen, pump and filter the water, keep the water within the required temperature range, make the chemical disinfectants added to the water, and to provide light and nutrition (Edwards and Watson 2011). Harvesting makes up 40% of the production costs. After that, shells, sand, and stones need to be removed, the seaweed milled into fine particles, and minerals removed that slow down the ethanol processes. Salt, chlorine, sulfur, nitrogen, iron, polyphenols, and metals can inhibit yeast fermentation. The next step is to transport your seaweed feedstock to the nearest seaweed ethanol maker, necessarily within 25 miles, or the transport cost will be too high. There, vast areas are needed to dry the seaweed in the sun, since it is too energy-intensive to heat up or centrifuge. It takes a square meter to dry just 3.5 dry ounces. The dried seaweed requires considerable pretreatment because algal polysaccharides are hard to ferment into ethanol (Milledge and Harvey 2016).

To be profitable, a seaweed ethanol plant needs to run year-round, but like land crops, the optimal harvest is once a year, and there is no known way to store kelp the rest of the year.

In the end, despite the far faster growth rate of seaweed than land plants, they are just not as good a source for ethanol. Corn grain is nearly three-quarters of carbohydrates that can be fermented into 463 L of ethanol per ton of raw material, but seaweed consists of just 6% fermentable carbohydrates, producing a paltry 38 L per ton (Aizawa et al. 2007).

Nor does seaweed scale up. In 2015, world production was only 30.4 million wet metric tons (Ferdouse et al. 2018), about four million tons after drying, far from the 15 billion tons of biomass needed to scale up enough for Europe alone (Richard 2010).

There is plenty of offshore areas to grow seaweed, and there are several experiments hoping to do this. In one, to avoid storms as well as destruction by passing vessels, drone boats can actually move floating seaweed farms, using long ropes to drag them out of harm's way, ultimately towing the farm back to shore for harvest. Since the surface of the ocean has few nutrients, the drones need to lower the seaweed up to 300 m (985 ft) at night (Ackerman 2017) and raise the farm in the day for sunshine.

Another pilot project, the Seaweed Paddock, will put a giant fence around free-floating Sargassum mats in the Gulf of Mexico. Then it will use wave-powered tug drones to tow its captive Sargassum to dead zones in the Gulf. Overloaded with nutrients, these dead zones will feed the Sargassum.

Other efforts to grow kelp by attaching it to offshore structures have not fared well. Waves have smashed seaweed farms into smithereens. Ironically, this is the same reason why dreams of commercial wave power have not borne out. The immense power of the waves tears these sturdy metal wave contraptions apart (DOE 2010; NRC 2013).

Maybe your tank will not be full of kelp. But the good news is your mouth can be filled with sushi and other foods made with seaweed (Csanyi 2019).

References

Ackerman E (2017) Robotic kelp farms promise an ocean full of carbon-neutral, low-cost energy. IEEE spectrum. https://spectrum.ieee.org/energywise/energy/renewables/robotic-kelp-farms-promise-an-ocean-full-of-carbon-neutral-low-cost-energy. Accessed 11 Nov 2020

Aizawa M, Asaoka K, Atsumi M, et al (2007) Seaweed Bioethanol Production in Japan – The Ocean Sunrise Project. OCEANS 2007, Vancouver, BC. https://doi.org/10.1109/OCEANS.2007.4449162

Chatman S (2018) Local gas station pumps water instead of fuel: consumers. NBCDFW. https://www.nbcdfw.com/news/local/local-gas-station-pumps-water-instead-of-fuel-consumers/237783/. Accessed 11 Nov 2020

Csanyi C (2019) What products are made from seaweed? Sciencing. https://sciencing.com/products-made-seaweed-7222937.html. Accessed 11 Nov 2020

DOE (2010) National algal biofuels technology roadmap. U.S. Department of Energy, Energy Efficiency and Renewable Energy, Washington, DC

Edwards M, Watson L (2011) Cultivating *Laminaria digitata*, vol 26. Aquaculture Explained; Bord Iascaigh Mhara (BIM), Dún Laoghaire, Ireland, pp 1–71

Ferdouse F, Holdt SL, Smith R et al (2018) The global status of seaweed production, trade and utilization. Food and Agriculture Organization of the United Nations, Rome

Milledge JJ, Harvey PJ (2016) Potential process 'hurdles' in the use of macroalgae as feedstock for biofuel production in the British Isles. J Chem Technol Biotechnol 91(2221):2234

NRC (2013) An Evaluation of the U.S. Department of Energy's Marine and Hydrokinetic Resource Assessments. National Research Council

Richard T (2010) Challenges in scaling up biofuels infrastructure. Science 329:793–796

Rogers-Bennett L, Catton CA (2019) Marine heat wave and multiple stressors tip bull kelp forest to sea urchin barrens. Sci Rep 9. https://doi.org/10.1038/s41598-019-51114-y

Chapter 27
The Problems with Cellulosic Ethanol Could Drive You to Drink

Corn ethanol is made from corn kernels, grain which is starch. Why not use the remainder of the plant—the leaves, roots, and stems of corn, or other plants altogether, like trees, and make cellulosic ethanol? This was a goal of the federal Renewable Fuel standard from the outset to avoid using food to produce ethanol. It was also assumed that plant-based ethanol would emit lower greenhouse gases than gasoline.

Despite being invented over a century ago and despite billions of dollars of investment, cellulosic ethanol never became commercial in the US, and the last commercial cellulosic ethanol plant closed in 2019 (Eller 2019).

Like kelp, plant bodies are extremely low fat, just 2–3%, so only cellulosic ethanol can be made. Not biodiesel to keep trucks, locomotives, and ships running.

Roots, stems, and leaves ought to be a great source of ethanol since they are chock full of fermentable sugars. But these sugars have proven really difficult to extract since lignin and compounds that hold plant cells together block access. To pry them out, scientists have mashed, radiated, exploded, and suffused plants with harsh chemicals. It is a good thing plants do not have feelings with all the plant abuse going on in this industry.

Although it appears cellulosic ethanol was finally made in quantity—in 2017 the EIA reported 10 million gallons produced—that is manure. Literally trash talk! The EIA changed the definition of cellulosic ethanol to include natural gas from manure and landfills (Ernsting 2016; BB 2017; EIA 2017; Ernsting and Smolker 2018).

What Plants Are Used to Make Ethanol in the US?

In addition to corn, a third of the sorghum crop is used to make ethanol.

Ethanol can be made from any plant with sugars or with starch that can be turned into sugar. But in that corn is mainly starch that must be converted to sugar, while

A. J. Friedemann, *Life after Fossil Fuels*, Lecture Notes in Energy 81, https://doi.org/10.1007/978-3-030-70335-6_27

beets already have lots of sugar, it beats me why ethanol is not made from beets. Beets are used in Europe and produce twice as much ethanol per acre as corn with just half the water. Unlike sugar cane, beets grow easily in a temperate climate. Did the corn lobby muzzle the sugar beet lobby? Beats me!

Corn kernels are 70% starch and easily broken down into glucose for fermentation. No other grain can be so easily converted into ethanol, because the corn kernel is uniquely vulnerable in the plant world, lacking a protective seed covering. Many plants with seeds evolved to depend on their fruit to be eaten since the seed inside can survive digestion with an impermeable outer layer, and … emerge, triumphantly, in a stinking pile of fertilizer far from the shade of the mother plant (Ray 2011).

Corn does not have a protective seed covering due to thousands of years of breeding and cultivation. It is now utterly dependent on us. Even a corn cob that falls on the ground will not reproduce since so many kernels compete with each other that none can grow. Corn needs humans to separate kernels from the cob and plant them a foot or more apart (Standage 2009).

Why Is Cellulosic Ethanol So Hard to Make?

Except for fruits and protected seeds, the rest of a plant evolved over hundreds of millions of years to *not* be eaten by herbivores or microbes, with barriers of toxins, spines, and thick bark. The most formidable defense is a rigid structure of indigestible cellulose, hemicellulose, and lignin, which even after death can take a year or more for microbes and fungi to consume and break down into new soil.

Scientists try to speed up the process with brute force. Bioreactors create high pressures and temperatures, other machines mill, radiate, steam explode, accelerate electrons, hydrolyze with acids, freeze, drench in harsh chemicals, expand fibers with ammonia or ozone, and inflict other torments to get the sugars out. Nothing much works. They have hit a cellulosic wall. The chemicals and tortures create toxins that can hinder or even kill the microbes needed to ferment the sugars for cellulosic ethanol, cutting production by half or more (Colin et al. 2011; NRC 2011, 2014).

Even when the many kinds of sugars in cellulose and hemicellulose are finally forced out, a single organism that can ferment all of them into ethanol does not exist. It can be done though. Insects and ruminant animals like cows have specialized digestive systems with many kinds of bacteria, fungi, and protozoa that digest leaves and stems over a few days.

Scientists have been trying for many years to replicate a termite's ability to break down plants. Termites digest wood by outsourcing the work to the protists in their gut. Protists, in turn, outsource the work to many bacteria that use enzymes to break wood down further. Just like at a factory, each microbe performs one task and excretes a different substance than it consumed. In a termite gut factory, one working microbes' poop is ambrosia for another. This intricate chain reaction has proven difficult to synthesize. Too much of anything along the chain of reactions and it can kill the process. For example, in ethanol production, when yeast has raised the concentration of excreted ethanol from 12 to 18%, the yeast dies.

So far scientists have not been able to get termite or ruminant gut organisms to expand from their tiny world into the expansive gut of a 2000-gallon stainless steel tank (Wald 2007).

There are scientists trying to make a genetically modified Swiss-army knife creature that can do it all: Snip, ferment, and excrete ethanol, but that has not happened yet either (NRC 2014).

The success of cellulosic ethanol depends on finding or engineering organisms that can tolerate extremely high concentrations of ethanol. Organisms have had a billion years of optimization through evolution to develop a tolerance to high ethanol levels. You would think generations of people making beer, wine, or moonshine would have discovered this creature if it existed.

Synthesizing termite gut and other enzymes is expensive. It can take up to 40 times more enzymes to break down cellulose than the enzymes used to fragment starch in corn kernels. Enzymes are the most expensive component of a cellulosic refinery and the main reason why cellulosic ethanol is not commercial (Sainz 2009).

Even if a cheap enzyme was discovered that is able to pry sugars out of corn stover lignin, there are so many varieties of corn that no single enzyme may be able to unlock the lignin from all of them (Sluiter et al. 2000; Ruth et al. 2013). Multiple enzymes may be required, just for corn stover. Processing a biomass batch that includes other plant material would increase the number of enzymes further.

In the end, enzymes are too slow, despite every effort of the biomass torture chamber to speed up the process (Fialka 2018). Too slow to become commercial.

EROI of Cellulosic Ethanol

You knew I was going to get you to consider the EROI of cellulosic ethanol. Nothing more fun than that!

Corn grain is harvested just once. But the remaining corn stover will need three more harvests using heavy equipment that may compact and erode the soil (Troeh and Thompson 2005). All three passes must be done before it rains or snows. The first pass—here we go, you know the drill—is done with a **windrowing flail shredder** that *VACUUMS UP* plant litter, *SHREDS* it to small pieces, and *BLOWS* it into long piles (windrows) of hay for baling. After drying, a **hay baler** *COMPRESSES* and *BALES* the residues. Finally, the bales are *HARVESTED* by a **bale loader-stacker**. A good farmer will have a fourth run to put some of the stovers back on the field to prevent erosion and add nutrition.

Then a **bale loader-stacker** *DRIVES* an average of 43 miles to the biorefinery is *UNLOADED* at the **storage area,** and the truck *RETURNS* empty (Perlack and Turhollow 2002). Each bale is *SENT* through a **machine** to remove the twine holding the bale together, then **giant scissors** *SHRED* the stover into smaller pieces and they are *SHAKEN* to remove dirt and rocks that can stop the process (Mayer 2017).

Then it is **GROUND DOWN** further to make as much surface area vulnerable to the enzymes, chemical treatments, and other methods that break the lignocellulose into sugars. Biomass is so tough and abrasive that these machines need to be larger and more powerful than in corn ethanol factories. Finally, biomass is **FED** into the digestive system of the ethanol plant, where many more steps—I am going to have mercy on you here—converts the pulverized plants into cellulosic ethanol.

Despite all this effort, about a quarter of the corn stover is lost during the harvesting and baling steps (Ruth et al. 2013), and an additional 6–54% if the bales get wet and compost (McGregor 2013).

The EROI of cellulosic ethanol is impossible to determine because it is not commercial yet, despite generous subsidies and tax credits. Given the many steps, it very likely has a negative energy return. Benemann et al. (2006) say it is easy to determine that it takes at least five times more hydrocarbon energy to make cellulosic ethanol than the energy produced. Pimentel and Patzek (2005) found that cellulosic ethanol from switchgrass required 50% more fossil energy and wood biomass 57% more fossil energy than the ethanol fuel produced.

Another Reason for Negative EROI: Plant Residues Are Fluffy

Corn stover is far less dense than corn grain. Corn grains weigh almost six times more than corn stover per cubic foot. Because of this, a cellulosic ethanol plant has a logistical nightmare to solve. Consider a plant set up to receive one million tons of stover during the 60-day harvest window. That plant would need to receive, store, and handle 1.6 million large bales weighing 1200 pounds with 32,000 trucks deliveries around the clock from thousands of square miles of farms (Brick and Lewis 2013).

The biomass storage area also must be huge, at least 100 acres to keep the biorefinery running until the next harvest (DOE 2003). Since hay can spontaneously combust and set the whole place on fire, or decay into compost if damp, the 100 acres of storage must be paved and have walls, vents to dissipate heat, and a roof to keep water out.

Energy Grass Crops Are No better than Food Crops

Though I have outlined a number of strikes against it, corn stover has a lot going for it. It is the favored choice for cellulosic ethanol production because it can be grabbed after the corn harvest, it is near existing ethanol factories, and there is a lot of it. But the more stover removed, the greater the erosion and fertilizer needed for next year's corn.

Straw from wheat and other grains will not do. The stalk is too short. The green revolution shrank the stalk to keep the plant from blowing over and put more energy into larger seeds. Soybeans do not have much residue either.

One way to get more cellulose is to grow energy crops meant only for cellulosic ethanol. Although it seems like a snap to grow millions of acres of weeds, energy crops are likely to be grown as monocrops because they are easier to manage and need fewer kinds of enzymes. Unfortunately, they will likely be just as attractive to pests as food crops are, requiring pesticides. And like crops, they will need water and deplete soil nutrition, so they will also need to be fertilized (McAloon et al. 2000; Nelson 2002; Sheehan et al. 2008).

Energy crops can be bio-invasive. Johnson grass was introduced as a forage grass and is now an invasive weed in many states. Wiedenmann describes *Miscanthus*, a hot cellulosic fuel candidate, as "Johnson grass on steroids" (Raghu et al. 2006; UARK 2006).

Wood is an option, but forests are not renewable unless you are willing to wait 50 years or more. While there are 132 million dry tons of milling residues, much of that goes to electricity generation in the US or Europe, where burning wood comprises half of all Europe's renewable energy consumption (Economist 2013; Cornwall 2017).

As a buyer of plant residues and wood, cellulosic ethanol makers may be outbid by other industries (Khanna and Paulson 2016). Alternative uses for plant residues include bioplastics, animal hay, furfural factories, hydro-mulching companies, bio-composite manufacturers, and pulp mills (NRC 2014).

Where Is the Land to Grow Energy Crops?

The federal Renewable Fuel Standard mandates that 16 billion gallons of cellulosic ethanol be produced annually. It might be a mandate, but it never happened. Today, we are 16 billion gallons shy of our quota. No cellulosic ethanol is being produced other than from manure and landfill gas.

Suffice to say, the alchemy for commercial cellulosic ethanol has yet to be created. Too, to meet the mandated 16-billion-gallon quota, 30–60 million acres (NRC 2011) would be needed. Land with adequate water and soil is already being used, not just for crops, but for pasture and rangeland that feeds 100 million cattle, five million sheep, nine million horses, and 2.6 million goats. This land also supports 30 million deer. Turn this land into cellulosic fuel farms and you will have a couple of very angry hunters on your doorstep.

Switchgrass is native to the Great Plains. Switchgrass is a darling of the cellulosic fuel crowd. Wu (2004) examined where it could be grown, focusing on northeast South Dakota. Researchers had a hard time finding places where cellulosic ethanol plants using switchgrass on Conservation Reserve program land could be located. All possible sites for a facility lacked enough water. Only eight sites met other requirements, which included: On a high-quality primary road or access to a railroad, adequate drainage; not in a floodplain; access to natural gas and electric power; on parcels of at least 40 acres to provide storage; not near towns, recreation, or wildlife areas; able to accommodate a wastewater disposal area; and sufficient land in the surrounding 72 square miles to grow at least 407 tons of cellulosic fuel stock. This will not be an easy row to hoe.

Conclusion

Plants have evolved over hundreds of millions of years to not be eaten. Cellulose, the main structural component of the primary cell wall of plants, protects plants because only ruminants and termites can digest it. It is a tough nut to crack. Despite hundreds of millions of dollars in research, partnerships between Big Oil, Stanford, and UC Berkley researchers, there is still no reason to think that unlocking the cellulose within plants to make ethanol can be solved in a few decades, or perhaps ever. If it can be done, the process almost certainly will have a negative EROI. And if cellulosic ethanol can not keep trucks running, then what is the point?

References

BB (2017) Corn fiber ethanol – examining 1.5G technologies. Biorefineries Blog. https://biorre-fineria.blogspot.com/2017/04/corn-fiber-ethanol-examining-1.5g-technologies-biorefineries.html. Accessed 11 Nov 2020

Benemann J, Augenstein DC, Wilhelm DJ, et al (2006) Ethanol from lignocellulosic biomass—a techno-economic assessment. Conf on Biofuels & Bioenergy, Univ. British Columbia, Vancouver, Canada http://www.rrapier.com/2006/08/guest-post-on-cellulosic-ethano/. Accessed 11 Nov 2020

Brick S, Lewis J (2013) Corn stover and the pace of cellulosic ethanol commercialization. Clean Air Task Force. https://www.catf.us/resource/corn-stover-and-the-pace-of-cellulosic-ethanol-commercialization/. Accessed 11 Nov 2020

Colin VL, Rodriguez A, Cristóbal HA (2011) The role of synthetic biology in the design of microbial cell factories for biofuel production. J Biomed Biotechnol. https://doi.org/10.1155/2011/601834

Cornwall W (2017) The burning question. Science 355:18–21

DOE (2003) Feedstock roadmap for agriculture biomass feedstock supply in the United States. U.S. Department of Energy, EERE

Economist (2013) Wood: the fuel of the future. Environmental lunacy in Europe. The Economist. https://www.economist.com/business/2013/04/06/the-fuel-of-the-future. Accessed 11 Nov 2020

EIA (2017) Renewable natural gas increasingly used to meet part of EPA's renewable fuel requirements. U.S. Energy Information Administration

Eller D (2019) Poet will 'pause production' at its Iowa plant that makes ethanol from corn cobs, husk. Des Moines Register. https://www.desmoinesregister.com/story/money/agriculture/2019/11/19/poet-pauses-production-ethanol-made-corn-cobs-husks-iowa/4233586002/. Accessed 11 Nov 2020

Ernsting A (2016) Subsidy loopholes for "cellulosic ethanol" promote corn profits, but not energy independence. https://inthesetimes.com/article/ethanol-subsidies-corn-biofuels-and-the-renewable-fuel-standard. Accessed 11 Nov 2020

Ernsting A, Smolker R (2018) Dead End Road: The false promises of cellulosic biofuels. Biofuelwatch. https://www.biofuelwatch.org.uk/wp-content/uploads/Cellulosic-biofuels-report-2.pdf. Accessed 11 Nov 2020

Fialka J (2018) How a government program to get ethanol from plants failed. Scientific American

Khanna M, Paulson N (2016) To harvest stover or not: Is it worth it? Farmdoc daily 6:32, Department of Agricultural and consumer economics, University of Illinois at Urbana-Champaign

Mayer A (2017) Cellulosic ethanol push stalls in the Midwest amid financial, technical challenges. https://will.illinois.edu/news/story/cellulosic-ethanol-push-stalls-in-the-midwest-amid-financial-technical-chal. Accessed 12 Nov 2020

McAloon A, Taylor F, Yee W, et al (2000) Determining the cost of producing ethanol from corn starch and lignocellulosic feedstocks. U.S. Department of Energy, National Renewable Energy Laboratory

McGregor J (2013) How rain affects hay quality and yield. Manitoba Co-operator. https://www.manitobacooperator.ca/livestock/how-rain-affects-hay-quality-and-yield/. Accessed 12 Nov 2020

Nelson RG (2002) Resource assessment and removal analysis for corn stover and wheat straw in the Eastern and Midwestern United States—rainfall and wind-induced soil erosion methodology. Biomass Bioenergy 22:349–363

NRC (2011) Renewable fuel standard: potential economic and environmental effects of U.S. biofuel policy. National Resource Council, National Academies Press

NRC (2014) The nexus of biofuels, climate change, and human health: Workshop summary. Institute of Medicine, National Resource council, National Academies Press

Perlack R, Turhollow AF (2002) Assessment of options for the collection, handling, and transport of corn stover. U.S. Department of Energy, Oak Ridge National Laboratory

Pimentel D, Patzek TW (2005) Ethanol production using corn, switchgrass, and wood. Biodiesel production using soybean and sunflower. Nat Resour Res 14:65–76

Raghu S, Anderson RC, Daehler CC et al (2006) Adding biofuels to the invasive species fire? Science 313:1742

Ray CC (2011) The toughest seed. New York Times. https://www.nytimes.com/2011/12/27/science/how-can-plant-seeds-survive-the-digestive-process.html. Accessed 12 Nov 2020

Ruth M, Mai T, Newes E et al (2013) Projected biomass utilization for fuels and power in a mature market. U.S. Department of Energy, National Renewable Energy Laboratory, Golden, CO

Sainz MB (2009) Commercial cellulosic ethanol: the role of plant-expressed enzymes. In Vitro Cell Dev-Pl 45:314–329

Sheehan J, Aden A, Paustian K et al (2008) Energy and environmental aspects of using corn stover for fuel ethanol. J Ind Ecol 7:117–146

Sluiter A, Hayward TK, Jurich CK, et al (2000) Compositional variability among corn stover samples. National Renewable Energy Laboratory. https://www.nrel.gov/docs/gen/fy03/33925.pdf

Standage T (2009) An edible history of humanity. Walker Books

Troeh F, Thompson LM (2005) Soils and soil fertility, 6th edn. Wiley-Blackwell

UARK (2006) Biofuels as invasive species? University of Arkansas. https://news.uark.edu/articles/9161/biofuels-as-invasive-species-. Accessed 12 Nov 2020

Wald ML (2007) Is Ethanol for the long haul? Scientific American. https://www.scientificamerican.com/article/is-ethanol-for-the-long-h/. Accessed 12 Nov 2020

Wu L (2004) Screening study for utilizing feedstocks grown on CRP lands in a biomass to ethanol production facility. National Renewable Energy Laboratory

Chapter 28
Biodiesel to Keep Trucks Running

Ever notice a car going down your street and smell French fries? Your nose is not lying. That vehicle is burning biodiesel made from reclaimed cooking oil.

We produce more than a billion gallons a year of biodiesel in the US. Our biodiesel is made from 95% vegetable oils (68% soybean, 16% corn, 11.4% canola) and 5% animal fats and grease (EIA 2019). About one-tenth of biodiesel comes from used cooking oil.

Biodiesel is the great hope, our main hope, to replace diesel for transportation and fossils in manufacturing.

As I have relentlessly hammered into you in previous chapters, nothing else can substitute, not ethanol from corn, cellulose, or kelp. Not biodiesel from algae, finite liquefied coal (CTL), hydrogen, ammonia, power-to-gas, methane hydrates, natural gas, or oil shale. Nor can transportation be electrified with batteries or overhead wires.

Biodiesel hits the ball out of the park. It is renewable. Trucks can run on it. It is commercial.

Although there are problems. You knew that was coming, didn't you?

Scale

Globally, 27.95 million barrels of petroleum diesel are consumed per day, but only 655,000 barrels of biodiesel (BP 2020). Biodiesel production would need to be scaled up 43-fold after oil decline.

Several previous chapters explained the myriad obstacles to growing more biomass. One more time, just for good time's sake, they include not enough land, soil erosion, water depletion, climate change, drought, and more.

The US burns 46 billion gallons of petroleum diesel a year. It makes just 1.8 billion gallons of biodiesel annually, 25 times less than needed (EIA 2018, 2019b).

A. J. Friedemann, *Life after Fossil Fuels*, Lecture Notes in Energy 81, https://doi.org/10.1007/978-3-030-70335-6_28

It bears repeating: 68% of US biodiesel comes from soybeans. Even if all 87.2 million acres of soybeans, grown on a quarter of America's cropland, were used to make biodiesel, just 5.7 billion gallons could be produced. But that is not likely, since soybeans are also in demand for livestock feed, cooking oil, baked goods, soy milk, tofu, industrial lubricants, and other goods (NCSPA 2019).

Corn can yield 18 gallons of biodiesel per acre. Soybeans can yield 65 gallons of biodiesel per acre.

What accounts for this differential? It is not crop yield. Corn yields 177 bushels per acre and soy just 52 bushels. Rather, it is fat content. Corn is 4% fat whereas soy is 20% fat. You need fat to produce biodiesel.

Despite its low-fat content (4%) and because of its high yield, corn manages to contribute 16% of annual US biodiesel production in 2019.

For biodiesel, peanuts would be better than either corn or soybeans. Peanuts are half oil and can yield 123 gallons per acre. But they grow only in the most humid areas of seven Southeastern states. Nor do other oilseed crops scale up.

Besides ramping up production, distribution will need to be scaled up too. There are 168,000 gas stations but only 300 where 20% biodiesel (B20) or above is available (AFDC 2020).

Biodiesel Requires a Lot of Water

The water footprint of biodiesel from seed to harvest to delivered fuel is vast, not good at a time of increasing drought, population growth, and aquifer depletion. Soy biodiesel requires 13,676 L (3613 gallons) of water per liter of biodiesel produced and corn ethanol 2570 L (725 gallons) (Gerbens-Leenes et al. 2009). Petroleum gasoline has a much smaller footprint, on average just 4.5 gallons of water per gallon of gasoline produced Wu and Xu (2018).

Water is a problem both coming and going. Plants can be up to 90–95% water. Removing this water to make fuels takes a lot of energy, but has to be done to avoid corroding and clogging diesel engines (Racor 2013).

Bad Chemistry

Making fuels from plants is challenging because their chemistry differs from crude oil, which is nearly all hydrocarbon chains of 82–87% carbon and 12–15% hydrogen.

Plants have hydrocarbons but are also chockablock with oxygen, nitrogen, phosphorus, potassium, calcium, magnesium, zinc, sulfur, chlorine, boron, iron, copper, manganese, and more. Good for vitamin pills, bad for trucks, these elements need to be removed to make biodiesel, adding to cost and energy.

Even then, plant oils are so different from petroleum diesel oil that it is hard to match the specifications of the diesel fuel standard. This standard, ASTM D 975, specifies energy density, oxidative and biological stability, lubricity, cold-weather performance, elastomer compatibility, corrosivity, emissions (regulated and unregulated), viscosity, cetane number, distillation curve, ignition quality, flash point, low-temperature heat release, metal, ash, and sulfur content, water tolerance, specific heat, latent heat, toxicity, and ash and sulfur content. It seems nearly miraculous that crude oil can be refined to conform to all these specifications. And it is a lot to ask of a soybean.

Unlike standard diesel, biodiesel is biodegradable and thus needs to be used within 45–90 days.

Why such extensive specifications for diesel? Fuel outside the specifications can harm diesel engines by gelling up in cold weather or acting as a solvent, releasing rust and other contaminants that plug filters and fuel injectors, and more (Bacha et al. 2007; Schmidt 2007).

That is why many heavy-duty engine manufacturers have warranties that do not allow biodiesel, though B5 (5% biodiesel/95% petroleum diesel) is often fine. Some warranties prohibit B20 to B100.

Distribution is a problem too. Biodiesel can not travel in oil or gas pipelines, because, like ethanol, biodiesel is a good solvent, and able to pick up water and impurities that can harm engines (APEC 2011). It is 5–20 times more costly to move biodiesel by rail or truck than were it possible to use pipelines (Curley 2008).

A Barrel of Crude Oil Is only 10–15% Diesel

It will take decades to build thousands of biodiesel factories, shift more cropland to oilseed crops, modify truck engines to burn B100, and build pipelines that can handle biodiesel.

If we could convert more crude oil to diesel, we could buy time for a transition from diesel to biodiesel. But only about 10–15% of a barrel of crude oil can be refined into diesel. A barrel of crude oil makes dozens of other useful products. As crude oil is heated at the refinery, fractions split off. The first to go are lighter hydrocarbons for plastics, then propane, gasoline, kerosene for jets, diesel, heavy oil, and asphalt.

The EROI of biodiesel is low, roughly 1.3–1.9 (Pimentel and Patzek 2005; Hill et al. 2006), far short of the 10–14:1 needed to keep civilization as we know it continuing (Lambert et al. 2014).

Conclusion

Biodiesel is significantly more expensive to make than petroleum diesel, so like ethanol, its existence is almost entirely due to federal policies such as the RFS biomass-based diesel and biodiesel production tax credits, excise tax credits, small biodiesel producer credits, and the RFS mandate that requires specific amounts of biodiesel in the overall fuel pool (Schnepf 2013).

Farmers can not grow enough oilseeds to replace petroleum diesel. But some of it is made from animal fats and grease. That brings us to the next question: Can we eat enough French fries to keep trucks running?

References

AFDC (2020) Biodiesel fueling station locations (B20 and above). U.S. Department of Energy, Energy Efficiency & Renewable Energy

APEC (2011) Biofuel transportation and distribution. Options for APEC economies. Asia-Pacific Economic Cooperation

Bacha J, Freel J, Gibbs A, et al (2007) Diesel fuels technical review. Chevron Corporation. https://www.chevron.com/-/media/chevron/operations/documents/diesel-fuel-tech-review.pdf. Accessed 12 Nov 2020

BP (2020) Statistical review of world energy 2020. British Petroleum

Curley M (2008) Can ethanol be transported in a multi-product pipeline? Pipeline Gas J 235:34

EIA (2018) Table 3.7c. Monthly energy review. Petroleum consumption: transportation and electric power sectors. U.S. Energy Information Administration

EIA (2019) Table 3. U.S. inputs to biodiesel production. U.S. Energy Information Administration

EIA (2019b) Table 10.4. Biodiesel and other renewable fuels overview. U.S. Energy Information Administration

Gerbens-Leenes W, Hoekstra AY, van der Meer TH (2009) The water footprint of bioenergy. Proc Natl Acad Sci 106:10219–10223

Hill J, Nelson E, Tilman D et al (2006) Environmental, economic, and energetic costs and benefits of biodiesel and ethanol biofuels. Proc Natl Acad Sci 103:11206–11210

Lambert JG, Hall CAS, Balogh S et al (2014) Energy, EROI and quality of life. Energy Policy 64:153–167

NCSPA (2019) Uses of soybeans. N.C. Soybean Producers Association. https://ncsoy.org/media-resources/uses-of-soybeans/. Accessed 12 Nov 2020

Pimentel D, Patzek TW (2005) Ethanol production using corn, switchgrass, and wood; biodiesel production using soybean and sunflower. Nat Resour Res 14:65–76

Racor (2013) Water, a diesel engine's worst enemy. Racornews. https://www.racornews.com/single-post/2013/12/05/Water-A-Diesel-Engines-Worst-Enemy. Accessed 12 Nov 2020

Schmidt CW (2007) Biodiesel: cultivating alternative fuels. Environ Health Perspect 115:86–91

Schnepf R (2013) Agriculture-based biofuels: overview and emerging issues. CRS Report R41282. Congressional Research Service

Wu M, Xu H (2018) Consumptive water Use in the production of ethanol and petroleum gasoline—2018 update. Argonne National Laboratory: Energy Systems Division. https://doi.org/10.2172/1490723

Chapter 29
Can We Eat Enough French Fries?

Gut check. If the US can not make enough biodiesel from plants, then the question becomes: Can we step up our fast-food game? Can we eat more French fries? Biodiesel is already made from used cooking oil (11.5% of all US biodiesel), animal, and other fats (5%). How about we Americans join up and every Monday night ply our tables and our bellies with fried food?

We could also scrap other products made from vegetable oils and animal fats and shift them to biodiesel. Products like soap, paint, varnish, vinyl plastics, lubricants, livestock and pet food, clothes, rubber, detergents, candles, rust inhibitor, shampoo, caulking, disinfectants, epoxies, electrical insulation, metal casting, plasticizers, and more (Murphy 2004). Who knew we are swimming in vegetable oils and animal fats!

Using restaurant waste grease helps stop the problem of it being illegally dumped into the sewage system, which can result in severe blockages. Once restaurant grease clogs a sewer, sewage may back up into the facility, shutting it down until the sewage pipe is cleaned. In London, sewers were clogged with a record-breaking fatberg that weighed 140 tons. Picture that!

Currently, hotels and restaurants generate three billion gallons of waste cooking oil, though most end up in landfills or down the drain rather than biodiesel (EPA 2017). That is a gob of grease, but a far cry from the 46 billion gallons of petroleum diesel consumed.

We will just have to belly up and eat a lot more deep-fried food. At last, we will have a good excuse to eat fried chicken and French fries with no guilt! If anyone can eat more fast food, it is Americans. Already 72% of adults are overweight or obese (CDC 2016). We are really good at eating. Spin eating fried food is a patriotic duty and it will be the most popular government edict ever.

Believe it or not, there are statistics on our personal grease consumption. About four gallons of used cooking oil and trap grease are generated per person per year (NREL 1998). Thank you for that important information NREL! With 329 million people, that is about 1.3 billion gallons. Americans are going to have to eat at least

A. J. Friedemann, *Life after Fossil Fuels*, Lecture Notes in Energy 81, https://doi.org/10.1007/978-3-030-70335-6_29

15 times more fast food to nudge biodiesel production to half of oil diesel consumption. Though obesity may shorten American lifespans, it will not be in vain. After death, excessive human fat can be rendered to make even more biodiesel to allow us to continue to our nonnegotiable way of life. Plus, liposuction clinics could harvest fat periodically. Americans could eat as much as they liked all the time, and we would forever end the pain of dieting (Squatriglia 2007).

It is too bad truck drivers running on empty can not just pull up to a fast food joint and say fill 'er up while eating fried food to keep the virtuous biodiesel cycle going. Sadly, unrefined vegetable oil, grease, or animal fats would harm their engine. Restaurant grease has to be trucked to a biodiesel refinery, pretreated to remove dirt, meat scraps, breading, and water, phospholipids, and plant matter degummed. Making biodiesel from cooking oil and trap grease is especially tricky because of their free fatty acids, which tend to react with catalysts to produce soap instead of biodiesel. The more soap made, the more soap and methanol that need to be removed and discarded, with some of the methanol and glycerol disposed of as hazardous waste. And to think this glob all began with French fries!

Conclusion

Seriously, Americans are not going to be able to eat enough fast food. Seriously, we are already too heavy. In an energy scarce world, it will be far better and healthier to live close to work and use muscle power, walking and bicycling, to get there. Seriously. Not that there will be any choice…

References

CDC (2016) Obesity and Overweight Adults. Centers for Disease Control and Prevention. https://www.cdc.gov/nchs/fastats/obesity-overweight.htm. Accessed 12 Nov 2020

EPA (2017) Learn about biodiesel. Environmental Protection Agency. https://19january2017snapshot.epa.gov/www3/region9/waste/biodiesel/questions.html#whyuse. Accessed 12 Nov 2020

Murphy DJ (2004) Plant lipids: biology, utilisation and manipulation. Wiley-Blackwell

NREL (1998) Urban waste grease resource Assessment. U.S. Department of Energy, National Renewable Energy Laboratory

Squatriglia C (2007) Around the world in a boat fueled by human fat. Wired. https://www.wired.com/2007/12/around-the-worl/. Accessed 12 Nov 2020

Chapter 30
Combustion: Burn Baby Burn

If oil is black gold, think of biomass as green gold. Biomass is a valuable energy commodity. We will never have enough of it. So, we better use it wisely. Burning it for heat is the most efficient use of biomass. When burned to generate electricity, it is only 20% efficient (Bain and Overend 2002). Likewise, it is far easier and more energy efficient to burn biomass for heat than to convert it to a liquid or gas.

Still, biomass is burned for electricity because electricity has more uses than heat. In terms of its use of water, the footprint for biomass generated electricity (bioelectricity) is better than biofuels.

Unlike wind and solar, which can generate electricity only when the wind or sun is out, biomass can generate electricity any time. On the other hand, a biomass plant can not fire up instantly, within seconds to minutes. That requires liquid or gaseous biofuels. So, it can not be used to balance the grid at those moments when wind and solar suddenly wane. Were biomass to be turned into a gas (biogas), then it could power plants that could fire up quickly to keep the grid in balance. But so little biogas is produced, and at a very slow rate, that it can not contribute much power to balance the grid (Arasto et al. 2017). Liquid biofuels could be used, but that would compete with more essential transportation and manufacturing.

In Europe, biomass, the fuel of preindustrial societies, accounts for half of the renewable energy, more than wind and solar, and is expected to grow to 60% of renewables in 2020. Biomass already accounts for over 80% of renewable energy in Poland and Finland (Economist 2013; Eurostat 2019). A substantial amount of the wood burned in Europe comes from the US (Elbein 2019). In the United States, wood contributes less energy because coal and natural gas still dominate, yet still, almost 1% of electricity is biomass generated (EIA 2019).

Burning biomass will not solve our energy crisis. To produce just 10% of US electricity (405 TWh) would require wood plantations the size of Minnesota to be cut down and chipped every year (Smil 2015).

A. J. Friedemann, *Life after Fossil Fuels*, Lecture Notes in Energy 81, https://doi.org/10.1007/978-3-030-70335-6_30

Burning Biomass is Dirtier than Coal

The outdated 2015 Paris climate change agreement counts burning biomass as carbon neutral. Pardon me, but I am going to join with the scientists that dispute that. Wood is dirtier than coal. Over 800 scientists have written a letter to the European parliament explaining that wood is not renewable and burning biomass will accelerate climate change by emitting 1.5 times as much CO_2 per kWh as coal and three times as much CO_2 as natural gas (Searchinger et al. 2018).

Even with pollution controls, a biomass plant will still emit 50% more carbon dioxide than a coal-burning plant, as well as twice as much nitrogen oxide, soot, carbon monoxide, and volatile organic matter (Upton 2014; Booth 2014). Forest wood emits antimony, arsenic, cadmium, chromium, copper, dioxins, furans, lead, manganese, mercury, nickel, polycyclic aromatic hydrocarbons, selenium, vanadium, and zinc. Treated wood is even worse (Biofuelwatch 2014).

Wood pellet factories and logging in the US Southeast and elsewhere are only profitable with government subsidies. Meanwhile, companies like Enviva have plans to grind a hundred square miles of forest in the Southeast into wood pellets for export to Europe and Asia.

Logging a forest—whether for lumber or for biomass—prevents nutrients like nitrogen and phosphate that are essential to new growth from being returned to the soil (Reijnders 2006; Biofuelwatch 2014). The next forest will consequently have fewer, smaller trees.

California Biomass Electricity Generation

California went big on BioRAMs. That is a biomass-burning plant. Half of US biomass power plants are in California (CBEA 2020). At one time, the state had 63 of them. Even with subsidies, 40 BioRAMs have gone out of business. The remaining 23 generate just 1.8% of California's electricity. About half of the biomass comes from agricultural and urban waste, another third from mill residues (sawdust), and 13% from forests.

More of these BioRAMs are likely to go out of business since electricity from wind, solar, natural gas, and nuclear power costs less. Tragically, a wildfire could be their salvation. In California, wildfires continue to burn hundreds of thousands of acres of forest year after year. In 2020, apocalyptic fires erupted. As of year-end, more than four million acres had burned: 4% of California's 100 million acres of land.

In an attempt to mitigate these catastrophic fires, several years ago California subsidized seven BioRAMs. For most of the year except winter, when roads are closed, they are powered in part by trees salvaged from burned areas and forests with a high danger of catastrophic fire, from High Hazard Zones (HHZ). The hope is that by reducing the fuel load in these areas, fires will be prevented or their scale reduced.

This seems like a good idea—generating electricity from dead trees that have been killed by wildfires, drought, and beetles. Even prior to the 2020 fires, there were 133 million dead trees in California.

When insects and pathogens attack trees, they ooze flammable resin for protection. After the tree dies, in a fire the resin will cause the flames to burn more intensely and spread more rapidly, and the dead needles enable out-of-control crown fires. But after a year or two needles drop, the resins are less volatile, and dead trees less combustible.

Just how much this taking of dead trees does to mitigate wildfire remains an open question. More certainly, dead trees, as they decompose, return nutrients to the ground, feeding the forest. Removing these natural time-released fertilizers impact the future of the forest.

Wood from Burned Forests to Generate Electricity

Despite the fires raging in California, Oregon, Washington, and Colorado as this book is being written in 2020, the cost to convert millions of acres of dead trees into electricity will remain steep.

Incinerating trees to make electricity is extremely inefficient, and heavily subsidized. In California, the levelized cost of biomass power averaged $166 per megawatt hour (MWh) compared to $49 per MWh for photovoltaic solar and $57 for wind (Neff 2019).

California's High Hazard Zone (HHZ) dead tree fuel costs three times more than non-qualifying fuel like sawdust, and is not available in the winter when forests are snowed in. HHZ wood costs a lot due to: $25 per bone dry ton to cut and skid the tree, $20 per ton to chip the HHZ wood, and $500 to haul 25 tons of chips to the BioRAM.

Other companies compete for HHZ material to make composite panels, bagged shavings, landscape materials, firewood, and logs and chips to export overseas.

HHZ forest wood has higher production costs than agricultural or urban waste wood to chip, load, and haul to a biomass plant. Transport is not economic if more than 50 miles, greatly limiting what can be burned. Of the 36 million acres of forests in California, about 16 million acres are in the HHZ zone, but only 3.6 million of those acres are within 50 miles of a BioRam plant. That was prior to the catastrophic widespread fires of 2020.

Although smaller plants could be built near burned or dying forests to reduce transportation costs, it is expensive to connect these new plants to the grid, and at least 100 of them would be needed in the Sierra Nevada alone (Quinton 2016).

Even when there is enough biomass to burn it has to be managed by moisture content and heating value, driving up handling and processing costs, limiting most biomass plants to about 50 MW (Rhyne et al. 2015). A 50 MW biomass plant is likely to cost substantially more per kW than a 500 MW coal-fired plant. These small biopower plants are less efficient than fossil fuel plants because it is too expensive to invest in high-efficiency technologies at such a small scale (NRC 2010).

Up to 80% of the HHZ biomass is uneconomical to haul and is left in the forest, with a small fraction burned in CalFire's dozen mechanized incinerators that burn dead trees onsite.

HHZ wood is chipped into small pieces in the forest before it is taken to the BioRAM because low-density branches would quickly fill up a truck with more air than wood. Chipper trucks grind the deadwood and load it into chip vans. Forest roads however are not designed for chip vans. Up to half of the chip vans can not negotiate roads designed for logging trucks with tighter turn ratios and higher ground clearance, leaving much potential HHZ wood on the forest floor. To make the roads accessible to chip vans would add up to $15 per ton of chips in additional costs.

The value of any wood salvageable as lumber from dead trees far surpasses the value of biomass for power generation. Either way, dead trees begin to decay quite quickly and must be harvested before deterioration. What makes HHZ biomass economic at all is that many of these dead trees can be salvaged for lumber, and the leftovers are given to BioRams for free. But due to new ways to salvage wood, less free wood is available.

HHZ wood is limited even more by the budget of the US Forest Service. As the climate heats up and forests burn, it is spending more on firefighting and less on fire management.

Biomass Fire and Explosion Hazards

Fire and explosion are serious threats to biomass storage, and these fires are difficult to extinguish. Conveyor belts are prone to dust buildup which can be ignited by frictional heating. Bucket elevators lifting bulk biomass can create flammable clouds of dust that can catch on fire and carry it to other parts of the plant (Ennis 2016).

Economic and Energy Costs (EROI)

HHZ biomass has 40% moisture on average, extra water weight that needs to be hauled to the BioRAM plant. On average a truckload of 25 wet tons delivers just 15 bone dry tons (BDT) for electricity production. California's goal is to burn a million BDT a year, which is 67,000 truckloads burning diesel fuel to and from the forest and BioRAM plant.

Getting forest biomass to businesses is complex to manage, expensive, and energy-intensive. There are a team of people involved, each with respective responsibilities:

1. **Forester**: Permits, develop contracts for the timber sale, logging price with sawmills; contracts for roads, logging, biomass, slash disposal, site prep, reforestation, order seedlings.
2. **Road contractor**: Construction, reconstruction, maintenance during the operation, road repair after the sale.
3. **Logging Contractor**: Get log haulers, cut, skid, and buck logs, load logs onto logging truck.
4. **Log Hauling contractor**: Haul logs to the sawmill.
5. **Biomass contractor**: Chip biomass and load chip vans.
6. **Chip hauling contractor**: Deliver chips to the biomass plant.
7. **Slash disposal**: Pile and burn logging slash.
8. **Site prep contractor**: Apply herbicide and do brush control.
9. **Reforestation:** Plant seedlings.
10. **Sawmill**: Receive logs and make lumber.
11. **Biomass plant**: Receive chips, make power.

With all these steps, turning dead trees into energy is not a freebie. No EROI studies have been done, but considering how many BioRAM plants have gone out of business and the subsidies, it would not be surprising to find the EROI was negative.

Conclusion

Bryce (2011) points out that "the problems with biomass-to-electricity schemes are the same ones that haunt nearly every renewable energy idea: Power density and energy density. Wood has only half the energy density of coal." And it is dispersed across the landscape, spread out over all tarnation.

A forest is a crop that takes a long time to grow. Half a century longer than corn. As fuel, it is slow and it is dirty. When burned, wood emits more CO_2 than fossil fuels (Dlouhy 2018).

Wood does not scale up. By 1900, even when there were only 75 million people, over half of North America was deforested (Williams 2006). Most of this was from wood use in our homes. In the eighteenth century, the average Northeastern family used 10–20 cords of wood per year. To sustainably harvest one cord of wood, at least one acre of woods is required (Whitney 1994).

Eleven percent of the homes in America burn wood for heat (EIA 2020). But if all of us used wood alone for heating and for nothing else, our country would be deforested in less than 33 years (Hagens 2007), especially in urban areas, where 80% of American's live, even in Montana (Swanson 2017). Which means that even in heavily forested states, or forests lacking roads, most of the trees will be too far away from the 80% in urban areas to harvest. State by state, here is how long the wood would heat its residents:

- Less than 1.2 years: California, Nevada, Arizona, South Dakota, Nebraska, Kansas, Illinois, New Jersey, Connecticut, Rhode Island, Massachusetts.
- Less than 1.6 years: Utah, Montana, North Dakota, Texas, Iowa, New York.
- Less than 2.5 years: Idaho, Wyoming, Colorado, Oklahoma, Indiana, Ohio, Maryland, Delaware.
- Less than 3.5 years: Pennsylvania, Michigan.
- Less than 5.5 years: Washington, Georgia, Maine, New Hampshire.
- Less than 9.5 years: Oregon, Alaska, Louisiana.
- Less than 16.5 years: Kentucky, Tennessee, Virginia, North Carolina, South Carolina.
- Less than 25–33 years: Arkansas, Hawaii, Mississippi, Alabama, Florida, West Virginia, Vermont.

I get the chills thinking about this. There is not enough biomass to keep us warm for long. And we will need wood for manufacturing, cooking, infrastructure, and other uses besides heating homes or electricity.

References

Arasto A, Chiaramonti D, Kiviluoma J, et al (2017) Bioenergy's role in balancing the electricity grid and providing storage options – an EU perspective. International Energy Agency

Bain RL, Overend RP (2002) Biomass for heat and power. For Prod J 52:12–19

Biofuelwatch (2014) Burning Wood in Power Stations: Public Health Impacts. http://www.biofuelwatch.org.uk/wp-content/uploads/Biomass-Air-Pollution-Briefing.pdf. Accessed 13 Nov 2020

Booth MS (2014) Trees, trash, and toxics: how biomass energy has become the new coal. Partnership for Policy Integrity. https://www.pfpi.net/trees-trash-and-toxics-how-biomass-energy-has-become-the-new-coal. Accessed 13 Nov 2020

Bryce R (2011) Power hungry: the myths of "green" energy and the real fuels of the future. Public Affairs

CBEA (2020) Biomass operations in California. http://www.calbiomass.org/facilities-map/. Accessed 13 Nov 2020

Dlouhy JA (2018) Trump embraces tree-fired power called worse than coal. Associated Press. https://apnews.com/article/491a0c526f264a128819e597d1f2eaae. Accessed 13 Nov 2020

Economist (2013) Wood: the fuel of the future. Environmental lunacy in Europe. The Economist. https://www.economist.com/business/2013/04/06/the-fuel-of-the-future. Accessed 11 Nov 2020

EIA (2019) Table 3.1.B. Net Generation from Renewable sources: total (all sectors), 2009-2019. U.S. Energy Information Administration

EIA (2020) Biomass explained. Wood and wood waste. U.S. Energy Information administration

Elbein S (2019) Europe's renewable energy policy is built on burning American trees. Vox.com https://www.vox.com/science-and-health/2019/3/4/18216045/renewable-energy-wood-pellets-biomass. Accessed 13 Nov 2020

Ennis T (2016) Fire and explosion hazards in the biomass industries. Hazards 26, Symposium series No 161. https://www.icheme.org/media/11801/hazards-26-paper-64-fire-and-explosion-hazards-in-the-biomass-industries.pdf Accessed 29 Nov 2020

Eurostat (2019) Wood as a source of energy. Eurostat. https://ec.europa.eu/eurostat/statistics-explained/index.php?title=Archive:Wood_as_a_source_of_energy&direction=next&oldid=427588 Accessed 9 Mar 2021

Hagens N (2007) Old sunlight vs ancient sunlight -an analysis of home heating and wood. Theoildrum.com. http://theoildrum.com/node/2683. Accessed 13 Nov 2020

Neff B (2019) Estimated cost of new utility-scale generation in California: 2018 update. CEC-200-2019-500. California Energy Commission

NRC (2010) Electricity from renewable resources: status, prospects, and impediments. National Research Council, National Academies Press

Quinton A (2016.) Biomass power could help California's dying forests. Valley Public Radio. https://www.kvpr.org/post/biomass-power-could-help-californias-dying-forests#stream/0. Accessed 13 Nov 2020

Reijnders L (2006) Conditions for the sustainability of biomass based fuel use. Energy Policy 34:863–876

Rhyne I, Klein J, Neff B (2015) Estimated cost of new renewable and fossil generation in California. California Energy Commission

Searchinger TD, Beringer T, Holtsmark B et al (2018) Europe's renewable energy directive poised to harm global forests. Nat Commun 9:3741

Smil V (2015) Power density. A key to understanding energy sources and uses. MIT press

Swanson (2017) Montana: one state with three changing regions (part 2 of 3). This is Montana, University of Montana. http://www.umt.edu/this-is-montana/columns/stories/montana_regions_2of3.php. Accessed 2 Dec 2020

Upton J (2014) What's worse than burning coal? Burning wood. Grist.org. https://grist.org/climate-energy/whats-worse-than-burning-coal-burning-wood/. Accessed 13 Nov 2020

Whitney GG (1994) From coastal wilderness to fruited plain: a history of environmental change in temperate North America from 1500 to the present. Cambridge University Press

Williams M (2006) Deforesting the earth: from prehistory to global crisis. University of Chicago Press

Chapter 31
Wood and Coal Steam Engines Started the Industrial Revolution

Before steam engines, the power to do work came mainly from water, wind, and muscles, especially horses, each able to do the work of six or more men.

The first commercial steam engine of the eighteenth century was a wonder to behold. Equal to five to ten horses in horsepower, it put the horse out to pasture. Steam engines do not need to rest, and since they can be located anywhere, they have advantages over wind or water power which must be sited in windy areas or near rivers.

It was not long before steam engines were doing the work of millions of horses, burning wood at first and then also coal. Steam engines produced unprecedented amounts of goods, powered locomotives, powered steamships, the bellows of blast furnaces, sawmills, water pumps, cotton gins, and threshers. For well more than a century, steam engines did the work that electricity and combustion engines do today.

Steam power is simple. Water is heated into steam, expanding up to 1700 times the volume of the water boiled, enabling the steam to drive pistons to do mechanical work.

Inventors discovered and began toying with the steam engine in 1698, but it did not become commercial until 1710. For 150 years, steam engines were the only type of fuel-burning machine. By the early 1900s, diesel and gasoline internal combustion engines had replaced many of them. As late as 1930, steam still powered nearly all trains and more than 80% of all ships (O'Connor and Cleveland 2014).

Steam engines were much larger and heavier than today's internal combustion engines. One behemoth, a 115-ton, five-story tall Corliss steam engine, pumped eight million gallons of water a day in New York (NIMH 2019). Today just four diesel pumps weighing a modest 5.4 tons can do that job (AWP 2020).

Although steam engines have disappeared, steam power lives on in steam turbines powered by coal, natural gas, nuclear power, oil, and concentrated solar power. They generate 80% of US electricity at an average 32.8% efficiency, with 67% of the potential energy wasted as heat (NRC 2015).

A. J. Friedemann, *Life after Fossil Fuels*, Lecture Notes in Energy 81,
https://doi.org/10.1007/978-3-030-70335-6_31

Steam Engines Won Wars

During the Civil War, the quartermaster in charge of goods for the Union army needed to provide 600,000 tons of supplies a day, including the food for a million horses and mules who consumed 28 pounds of hay and corn a day. A six-mule team can haul up to two tons of goods a day for seven miles on decent roads, so imagine the number of mule-teams required and the time it took to haul animal feed and all the other goods over mostly terrible roads. It was not long before the industrialized north ramped up their use of steamships and locomotives to quickly move men, horses, and supplies for hundreds of miles. Steam engines displacing horses, mules, and sailing ships made this the first modern war (SAAM 2015; O'Harrow 2016).

Noisy, Complex, and Lots of Maintenance

Stationary steam power in factories was not altogether stationary. The steam-powered large webs of leather belts, pulleys, gears, and line shafts were noisy, vibrating, very dangerous, and got in the way of lighting, overhead cranes, and ventilation ducts. All of these components broke down and needed maintenance. And once a week, some poor soul at the factory got the job of cleaning out the boiler.

For a long time, locomotives used wood steam power. In the 1880s, skidders hauled cut trees out of forests onto railroad cars. Skidders were strange beasts that looked like a locomotive with a ship's mast with long cables extending every which way. Each cable was attached to a fallen tree that was skidded hundreds of feet to the rail tracks for loading. Nearly everything in the nearby forests was cut, leaving just stumps, branches, and bark that burned when sparks from a railroad wheel or glowing ash from a tinderbox fell, at times starting fires that burned through the cutover hills to forests beyond (Stoll 2017).

After railroads entered the Midwest some of the worst forest fire disasters in American history began, fires then unparalleled in size and intensity, consuming millions of acres. The firestorms began in 1870 and did not end until the timber was exhausted in the 1930s and agriculture was forbidden on millions of acres. The conflagrations came from a combination of massive logging to power locomotives and clearance for farms that left behind up to 15-foot-high mountains of flammable slash. Farmers set fires to clear the land for crops. The skies grew so dark that noon became night, fires roared as loud as a freight train, and when the flames arrived, they appeared to drop from the sky like a hurricane, with winds that could toss wagons. Farms and fires changed the north woods from pines to mostly noncommercial tree species, and in some areas vaporized soil cover down to the rock layer below (Pyne 2017).

Locomotives were difficult to maintain and operate. Across the west, there was often bad water, with a high mineral content or alkali that could foul the boiler, requiring train companies to build and operate water treatment plants, another cost added to the expense of cleaning the mineral deposits out of boilers.

Locomotives were available just a third of the time due to a high amount of maintenance (Nilsson 2013). Today's diesel-electric locomotives are available 95% of the time (Hoffrichter et al. 2012).

Before they burned coal, steamships relied on thousands of wood yards along the banks, with timber hacked to a size that ships' furnaces could use. At night the owners of these wood lots kept large fires blazing so that steamships could find them in the dark. Steamships needed to stop every 2 h to fuel up, and several hours more to be loaded (Perlin 2005). At best, with incessant stoking, steamships could make 9 mph. Today's container ships move at a speed of 27 mph (Smil 2013).

Wood is Less Energy-dense than Coal or Oil

Wood is both less energy-dense and more dispersed than fossil fuels. Steam engines burned recently cut green wood with a high moisture content to stoke steam engines. The typical moisture content of freshly cut hardwoods is about 60% (Bond and Downing 2013). Burning wet wood reduces the energy output of a steam engine. The people who operated these steam engines at times elected to kiln dry the wood, or to air dry it. One process takes energy and the other takes time.

Whenever bituminous coal could be obtained, it was the preferred fuel, with 10–14,000 BTU/lb. Coal takes up just half the volume of an equal weight of wood.

Railroad companies routed their tracks past coal mines for refueling and for hauling coal elsewhere. Even in 2014, 40% of the tonnage hauled by trains was coal. But by 2019, as natural gas replaced coal, only 14% of rail tonnage was coal (AAR 2020a).

Since coal can heat much larger volumes of water much faster than wood, coal-driven steam engines can be much larger than wood steam engines.

Coal beats wood. Oil beats coal. Dense with energy, oil provides 18,300 BTU/lb. And rather than needing to be shoveled or tossed, it can be pumped through pipes. But you burn what you have. In that California had almost no coal, but lots of forests, most of the forests in the state would have been burned in steam engines were it not for oil and imported coal.

Enormous Amounts of Fuel and Water were Needed

Back in the day of the coal-powered rail lines, each week a locomotive consumed its own weight in coal and water. One of the largest locomotives, the UP Big Boy 4014, weighed 1.2 million pounds. Every week it needed 144,000 gallons of water, a fifth of an Olympic sized swimming pool.

If this locomotive had burned wood rather than coal, it would have chewed through 300 cords of wood every week, enough to fill up 11 40-foot transportation containers.

Energy Efficiency and EROI

Before a locomotive burning coal could get going, a fire had to be built in order to get the boiler up to a high enough temperature. The water level needed to be maintained and fire constantly stoked with new shoveled coal, and airflow maintained by opening and closing vents. This wore out the parts which needed to be replaced often.

Steam locomotives had a very low thermal efficiency. They used a vast amount of energy to build up steam pressure, which had to be discarded whenever the locomotive stopped or shut down.

Steam locomotives were about 4–8% efficient on average, so 92–96% of the energy in the wood or coal was wasted in the boiler (Ayres et al. 2003; Heck 2011). Locomotive speed and efficiency were also limited by their very heavy-weight and boiler capacity (Krug 2019). At best steam locomotives achieved 9–13% efficiency.

The best stationary steam engines may have been 11–18% efficient (Ayres et al. 2003; ANL 2012).

Compare that to a diesel-electric locomotive that can be over 40% energy efficient and consequently go further on far less fuel (Stodolsky 2002; Hoffrichter et al. 2012). Even gasoline internal combustion engines can be 25% efficient.

No wonder steam engines ultimately were replaced with combustion engines. And not just for efficiency. Steam engines were far too heavy to compete with internal combustion engines, which weigh less than 1/1000 of what early steam engines weighed (Smil 2017).

The steam locomotives of the Southern Railway from 1910 to 1955 could haul 700–1000 tons for 150 miles on 14,000 gallons of water and 16 tons of bituminous coal. Modern freight trains can move one ton of freight 470 miles on a single gallon of fuel, refueling far less often than steam locomotives (AAR 2020b).

Deforestation

Deforestation has been the ruin of many civilizations (Perlin 2005).

By the 1600s, fuel shortages began to occur in England. Forests had been cut to the point where few woodlots over 20 acres remained. In southeastern England, there was virtually no fuelwood at all. Luckily for England, there was plenty of coal to replace wood.

In 1850, 25% of the land area of the United States was densely forested; just 20 years later, this figure had dropped to 15%. By 1867 almost five billion cords of wood had been consumed for fuel in fireplaces, industrial furnaces, steamboats, and railroads. This erased about 200,000 square miles of woodlands, an area the size of the states of Illinois, Michigan, Ohio, and Wisconsin. Half of all these cords were consumed in just 17 years, from 1850 to 1867 (Perlin 2005).

In the US, people cut forests down to clear land for crops, heat buildings, burn wood in locomotives, steamships, and factory steam engines, and for the charcoal to make iron (14 tons of charcoal to smelt one ton of iron ore). At the start of the eighteenth century, 85% of Massachusetts was forested, but only 30% by 1870, when there were only 1.5 million people (Smil 2017). In West Virginia, the forests were almost wiped out in only 50 years. Of the state's 10 million acres of virgin forest in 1870, only 1.5 million remained in 1910 (Stoll 2017).

And yet, somehow, we still have forests. They survive because coal and oil replaced wood. By 1884, coal and oil exceeded biomass as sources of energy (Smil 2017). I do not want to imagine a planet without forests.

Boiler Explosions

Due to extreme pressure—steam can expand 1700 times the volume of water—steam engines can burst. In the early days of steam engines, the boilers were made with materials that could not dependably handle high steam pressure because the iron and steel were made with charcoal, which often does not reach the high temperatures needed to soften metals. This can be catastrophic. The worst boiler explosion happened in 1865 on the steamboat Sultana when 1549 passengers were killed near Memphis, Tennessee.

Steam Power will Come Back to Replace Muscle Power

Despite all the disadvantages of steam engines and their low efficiency, steam power will likely come back after fossils are gone. Mechanical power will be keenly desired considering the alternative—muscle power.

An average human working for 8 h can generate about 100 watts of power, by say, pedaling a stationary bike that generates electricity. That would be enough to watch TV. You could watch while pedaling. Hey, no slacking off! Keep spinning, the show has another 30 min to go.

People are actually looking into this! A study investigated how much electric power could be generated by 40 people at a gym in South Carolina. At best, 5% of the gym's average daily electricity demand could be provided. In terms of the value of the electricity produced, the cost to pay back the modifications to the rowing machines to generate electricity would take 33 years (Carbajales-Dale and Douglass 2018). Have you ever seen a 33-year-old rowing machine?

Energy Slaves

Fossil fuel-driven machines are the equivalent of modern energy slaves. Hagens (2020) estimates that every barrel of oil represents 4.5 years of human labor. Rickover and Admiral (1957) estimated that a locomotive was the equivalent of 100,000 men, a jet 700,000 men, and at least 2000 men push each automobile along the road.

Overall, the oil the world burns every year represents the energy equivalent of 66 billion energy slaves (Alexander 2012). Perhaps even more—if a barrel equals 4.5 years of labor, then the 34.7 billion barrels of oil extracted in 2019 is equal to 156 billion energy slaves. Add in billions more coal and natural gas slaves and you have got quite an army. Many people have speculated that this fossil fuel windfall is the main reason we abandoned slavery (Cobb 2005). Though tragically, there are still roughly 30 million slaves today (Fisher 2013).

Buckminster Fuller was one of the first to propose the concept of energy slaves, who "are enormously more effective because they can work under conditions intolerable to man, e.g., 5,000° F, no sleep, ten-thousandths of an inch tolerance, one million times magnification, 400,000 pounds per square inch pressure, 186,000 miles per second alacrity and so forth." (Fuller 1963).

In 1945, farms and ranches employed 25 million people, 7.5% of America's population. Today less than a million people work our farms and ranches, far less than 1% of the country. That is, thanks to diesel-powered tractors, combines, balers, seeders, generators, milking machines, shredders, irrigation pumps, and other farm equipment. No wonder farmers love their machines.

Steam Engines were the First Form of Energy Able to Reproduce Itself

Steam will rise once again as an energy source. After fossils decline, steam engines mainly will be used in factories and construction. Our descendants likely will run locomotives and site ports near forests and remaining coal deposits whenever it is possible to do so. Just as likely, forests will be overcut and disappear on these corridors.

Technologically it is doable. The steam engine is very simple and requires only basic skills to produce and maintain. Blacksmiths can forge scrap steel into boilers, pistons, and other parts.

Post fossils, the need for precision tools still will be needed, and steam power can be used to make machine tools, which can cut, grind, bore, and shear metals. Machine tools include lathes, milling machines, and pillar drills. These machines can be precise to one ten-thousandth of an inch to make interchangeable parts for clocks, sewing machines, screws, locks, and pistons for steam engines, and more (Vogel 2003; Winchester 2018).

Some of today's technology is precise to over a millionth of an inch, e.g., computer chips and airplane turbines. This level of technology will be hard to maintain after fossil fuels since metal alloys are required that require more heat than wood charcoal can deliver.

Horses will be Recalled from the Pasture

Horses, oxen, and mules will someday play the same roles they did in the past. On the other hand, steam tractors will not work. Unavoidably heavy, they compact the soil, get stuck in the mud, need to be near large supplies of water and wood, and are hard to move to other farms since they can be too heavy for bridges.

For similar reasons, steam-powered trucks are not likely either since they would destroy remaining roads with their immense weight and be hard to fuel with thousands of stacks of wood and barrels of water along roads.

Any society that attempts the return to steam will quickly deplete its woodlands, which are already succumbing to wildfires and climate change. As it is, wood will be in short supply as the main source of infrastructure and the only way to heat, cook, smelt metals and make bricks, glass, and ceramics. As the forests disappear, society would wither. We would have run out of steam.

References

AAR (2020a) What Railroads Haul: coal. Association of American Railroads. https://www.aar.org/wp-content/uploads/2020/07/AAR-Coal-Fact-Sheet.pdf. Accessed 2 Dec 2020

AAR (2020b) Freight rail: moving miles ahead on sustainability. Association of American Railroads. https://www.aar.org/article/freight-rail-moving-miles-ahead-on-sustainability/#! Accessed 14 Nov 2020

Alexander S (2012) Peak oil, energy descent, and the fate of consumerism. University of Melbourne – Office for Environmental Programs. https://doi.org/10.2139/ssrn.1985677. Accessed 14 Nov 2020

ANL (2012) Advanced vehicle technologies. Outlook for electrics, internal combustion, and alternate fuels. U.S. Department of Energy, Argonne National Laboratory

AWP (2020) MTP4DZV Magnum. Absolute Water Pumps. https://www.absolutewaterpumps.com/magnum-by-generac-mobile-dry-prime-diesel-trash-pump-mtp4dzv-4-inch-1450-gpm-deutz-diesel-skid-or-trailer-mounted. Accessed 14 Nov 2020

Ayres RU, Ayres LW, Warr B (2003) Exergy, power and work in the US economy 1900-1998. Energy 28:219–273

Bond B, Downing A (2013) Firewood, facts, follies and forest management. Virginia Tech college of natural resources and environment. https://forestupdate.frec.vt.edu/content/dam/forestupdate_frec_vt_edu/resources/presentations/WoodsandWildlife2013/bondfirewood.pdf. Accessed 14 Nov 2020

Carbajales-Dale M, Douglass B (2018) Human powered electricity generation as a renewable resource. Biophys Econ Resour Qual 3. https://doi.org/10.1007/s41247-018-0036-5

Cobb K (2005) Can democracy survive without fossil fuels? Resource insights. http://resource-insights.blogspot.com/2005/06/can-democracy-survive-without-fossil.html. Accessed 14 Nov 2020

Fisher M (2013) This map shows where the world's 30 million slaves live. There are 60,000 in the U.S. The Washington Post

Fuller RB (1963) World design science decade 1965-1975. Inventory of world resources, human trends and needs. Southern Illinois University, Carbondale, IL. https://www.bfi.org/design-science/primer/world-design-science-decade. Accessed 14 Nov 2020

Hagens NJ (2020) Economics for the future -- Beyond the superorganism. Ecol Econ 169

Heck RCH (2011) The Steam Engine and Turbine - A Text Book for Engineering Colleges. Read Books Limited

Hoffrichter A, Miller AR, Hillmansen S et al (2012) Well-to-wheel analysis for electric, diesel and hydrogen traction for railways. Transp Res Part D: Transp Environ 17:28–34

Krug A (2019). Steam vs diesel. The railway technical website. http://www.railway-technical.com/trains/steam-vs-diesel.html. Accessed 14 Nov 2020

Nilsson J (2013) Why you don't see steam locomotives anymore. Saturday Evening Post. https://www.saturdayeveningpost.com/2013/05/locomotive-diesel-engine/. Accessed 14 Nov 2020

NIMH (2019) Machinery hall. National Museum of Industrial History. https://www.nmih.org/corliss-weekend/. Accessed 14 Nov 2020

NRC (2015) Overcoming barriers to deployment of plug-in electric vehicles. National Academies Press, Washington, DC

O'Connor PA, Cleveland CJ (2014) U.S. Energy transitions 1780-2010. Energies 7:7955–7993

O'Harrow R (2016) The Quartermaster: Montgomery C. Meigs, Lincoln's General, Master Builder of the Union Army. Simon & Schuster

Perlin J (2005) A Forest journey: the story of wood and civilization. Countryman Press

Pyne SJ (2017) Fire in America: a cultural history of Wildland and Rural Fire. University of Washington Press

Rickover HG, Admiral (1957) U.S. Navy. Energy Resources and Our Future. Scientific Assembly of the Minnesota State Medical Association http://large.stanford.edu/courses/2011/ph240/klein1/docs/rickover.pdf. Accessed 14 Nov 2020

SAAM. 2015. How the railroad won the war. Smithsonian American Art Museum. https://americanexperience.si.edu/wp-content/uploads/2015/02/How-the-Railroad-Won-the-War.pdf. Accessed 14 Nov 2020

Smil V (2013) Prime movers of globalization: the history and impact of diesel engines and gas turbines. The MIT Press

Smil V (2017) Energy and civilization a history. The MIT Press

Stodolsky F (2002) Railroad and locomotive technology roadmap. U.S. Department of Energy, Argonne National Laboratory

Stoll S (2017) Ramp hollow: the ordeal of Appalachia. Hill and Wang

Vogel S (2003) Prime mover: a natural history of muscle. W. W. Norton

Winchester S (2018) The Perfectionists: How Precision Engineers Created the Modern World. HarperCollins

Chapter 32
Wood Gas Generators

Although future generations will likely build steam engines, especially stationary steam engines for manufacturing, they are very inefficient. They average just 4–8% efficiency, with the very best locomotive engine achieving 11% in 1900 (Ayres et al. 2003). Like biofuels, steam engines will be limited by how much wood or coal is available. Local rail is more likely than continental rail given how difficult it would be to distribute bone dry wood piles along a rail route or to take along enough rail cars of wood or coal.

Wood gas generators offer another potential option in regions with plentiful wood. A wood gas generator is an onboard gasification unit that converts wood or charcoal into wood gas that can then be used to power an internal combustion engine. During World War II, onboard wood gas generators were used to provide fuel for over a million cars, small trucks, buses, boats, and trains. Imagine having a wood gas generator in the trunk of your car!

Mechanics have told me that wood gas generators are impractical because of the large amount of wood it takes to produce a small amount of gas. Nor is it likely that newer cars could run on wood gas without reprogramming the car's computer and several other sensors that adjust fuel, air, and combustion amounts and timing (Hincke 2020).

We have not lost the knowledge of how to make wood gas generators. Detailed information on how to engineer them exists along with images of these amazing contraptions (LaFontaine and Zimmerman 1989; FAO 1986). They can be made from metal garbage cans, washing machine tubs, water heaters, and they could probably do double duty, making moonshine as well. In that they need to hold as much wood as possible, they weigh hundreds of pounds and take up most of the trunk of an auto. You would have not only a gas plant in your trunk but a pile of wood in your back seat.

So, let us talk—I would like to sell you a car. A wood-fueled car. I will throw in a cord of wood. And yes, I am required to make full disclosure about what you are buying. Firstly, it takes a while to fire up one of these vehicles. So, you will get a

© The Author(s), under exclusive license to Springer Nature Switzerland
AG 2021
A. J. Friedemann, *Life after Fossil Fuels*, Lecture Notes in Energy 81,
https://doi.org/10.1007/978-3-030-70335-6_32

complimentary 10-min break while the system heats up enough to travel. And you will love taking care of your vehicle. For every hour of driving time, you will relish the 20 min or so of hands-on attention to refuel, drain the condensate tank, clean the hose, fabric filter, condensate jacket, and best of all, remove the ash (FAO 1986).

You will really enjoy feeding your smoking hot beast. It likes its wood chopped into small bite-sized blocks about 3 inches long and two inches wide. Logs are too large, sawdust too small, and twigs, bark, and sticks unacceptable. And did I mention that it prefers wood with a moisture content of less than 20%, and aged for at least 6 months. Very discriminating, yes?

About two pounds of wood will take you an amazing one mile down the road (Peterson 2020). A manual for building a gasifier estimated an average truck would burn 3.5 pounds of wood a mile and strongly advised against running out of wood. Filling up is not as simple as cutting down the nearest tree. The owner of that tree might not like that. Not to mention that freshly cut timber has a 50% moisture content—too wet for the gasifier (Colmant 1939). So, keep your eye on that fuel gauge!

I think I have closed the sale. Will that be cash or would you like a loan?

North Korea Shows the Way

Today, the only wood-powered gasifiers in use are in North Korea, where fossil fuels are extremely scarce due to sanctions. About 70% of vehicles are military. Filling 'er up is no problem. The military is empowered to gather whatever wood they like, driving up the price and scarcity of wood for everyone else (Wogan 2013; RFA 2016). Hey, that is my garage you are chopping down!

Elsewhere, we are not seeing a lot of wood-powered gasifiers. No surprise here: LaFontaine and Zimmerman (1989) warn that "wood gas generators are not technological marvels that can totally eliminate our current dependence on oil, reduce the impacts of an energy crunch, or produce long-term economic relief from high fossil fuel prices, but they are a proven emergency solution when such fuels become unobtainable in case of petroleum shortages, war, civil upheaval, or natural disaster."

References

Ayres RU, Ayres LW, Warr B (2003) Exergy, power and work in the US economy 1900-1998. Energy 28:219–273
Colmant A (1939) Manuel Pratique des Automobiles a Gazogène. Chiron Etienne
FAO (1986) Wood gas as engine fuel. FAO forestry paper 72. Food & agriculture organization of the United Nations
Hincke C (2020) Private communication
LaFontaine H, Zimmerman FP (1989) Construction of a simplified wood gas generator for fueling internal combustion engines in a petroleum emergency. U.S. Federal Emergency Management Agency, Washington, DC

Peterson B (2020) Wood gasifier builder's bible: off grid fuel for the prepared homestead: wood gas in minutes. Independently Published

RFA (2016) Charcoal powered vehicles stage a comeback in North Korea. Radio Free Asia. https://www.rfa.org/english/news/korea/charcoal-powered-vehicles-make-a-comeback-in-north-korea-12092016160533.html. Accessed 14 Nov 2020

Wogan D (2013) How North Korea fuels its military trucks with trees. Sci Am https://blogs.scientificamerican.com/plugged-in/how-north-korea-fuels-its-military-trucks-with-trees/. Accessed 14 Nov 2020

Chapter 33
Conclusion: Do You Want to Eat, Drink, or Drive?

As seen in the dark of night from space, the illuminated cities and population centers of the continental United States. The image is a composite from satellite data (NASA 2012).

"My grandfather rode a camel, my father rode a camel, I drive a Mercedes, my son drives a Land Rover, his son will drive a Land Rover, but his son will ride a camel."
—Rashid bin Saeed Al Maktoum.

© The Author(s), under exclusive license to Springer Nature Switzerland AG 2021
A. J. Friedemann, *Life after Fossil Fuels*, Lecture Notes in Energy 81,
https://doi.org/10.1007/978-3-030-70335-6_33

This book essentially is a reality check of where energy will come from in the future. It has not been an easy read. Nevertheless, we persist. So, let us review.

The electric grid can not stay up without natural gas due to a lack of energy storage. Transportation, agriculture, and manufacturing can not be electrified. If transportation can not be electrified or run on Something Else, civilization as we know it ends.

Agriculture goes back to horses and human labor. When trucks stop running, manufacturing will only exist at ports or remaining forests. With wood charcoal as the source of high heat and biomass as the source of many goods, manufacturers will make far fewer and much lower quality goods.

Youngquist (1997) wrote that the destiny of nations depended on their mineral and fossil fuel resources. The geodestiny of the US was blessed with unmatched mineral and energy resources and went from a wilderness to the most rich and powerful nation on earth. Today each person in the US uses over 3.19 million pounds of minerals, metals, and fuels in their lifetime (MEC 2020). But we have exploited most of our minerals, and now need to import 50–100% of 49 minerals and all 17 rare earth metals essential for computers, cell phones, cars, and other high-tech (Reuters 2019; USGS 2020).

We are running out of time, energy, and mineral resources to replace fossil fuels, despite having had all of human history and the last few centuries to find Something Else. Energy transitions take decades. It took 50 years for oil to capture 10% of global energy after it was first drilled in the 1860s, and 30 more years to provide 25% of all energy. It took 70 years for natural gas to go from 1% to 20% of global energy (Smil 2010).

The larger the scale of existing infrastructure, the longer fossil substitution will take. In 2019, wind and solar contributed just 1.3% of total world energy consumption (BP 2020).

Ethanol was first used in combustion engines in 1826. Rudolph Diesel invented the diesel engine in 1890 with the intention of running it on biological fuel. The first practical battery, the Daniell cell, was invented in 1836. The first hydrogen fuel cell was invented in 1839. The energy crises in the 1970s led to the Department of Energy being established in 1977, and since then billions of dollars have funded university and national lab research on energy.

The basic and unsolved problem is that alternative sources of energy require fossil fuels for every step of their life cycle. The wind may be free but it takes a lot of fossil energy to create windmills. Sunlight may rain down on the Earth but you can not make photovoltaics without fossil energy. Even if you are willing to force a positive EROI by using narrow boundaries that overlook many fossil inputs, their EROI is still nowhere near the 10:1 required for our civilization.

None of the alternatives produce enough energy to "reproduce."

As energy expert Vaclav Smil wrote, scaling biomass production to supply a significant share of the world's liquid biofuels is delusionary (Smil 2017). Let alone enough biofuels or biomass charcoal to make cement, steel, glass, bricks, plastics, or ceramics; nor electricity, the feedstock for 500,000 products, heating, cooking, fertilizer, and so on.

Nor can we just grow more plants, there is not enough land. Pesticides will stop working eventually. Crop productivity will drop from topsoil erosion, water depletion, and climate change at a time when three billion additional hungry mouths are on the way.

Biofuel energy return is negative or barely positive when made with higher energy-dense fossil fuels, so it will certainly be negative when biodiesel is used to make biodiesel.

Nonetheless, we should take what we know can do with plants and run with it, plan for such a world, kicking and screaming all the way, but dealing with reality.

We are nearing the end of a one-time binge. We flew high above oceans and mountains, drove trillions of miles, and flipped switches for light, TV, and air-conditioning. We live like Gods compared to all of past human history.

But as oil declines, civilization will revert to biomass for thermal energy, as well as muscle, river, and wind power just as in the past.

Let us look on the bright side! Honestly, there is an upside. As oil declines, the threats of hothouse earth and extinction from climate change decline.

Climate models developed by the Intergovernmental Panel on Climate Change (IPCC) show a range of greenhouse gas trajectories. The worst-case IPCC scenario of Representative Concentration Pathway (RCP) 8.5 with a temperature increase of 5 °C would lead to a hothouse world. The media, policymakers, and experts often depict this outcome as the most probable "business as usual" future. But it was meant to depict an unlikely future. Some scientists think around 3 °C (RCP 4.5 to RCP 6) is more likely (Hausfather and Peters 2020).

On the other hand, geologists using climate models with realistic, more limited global fossil fuel reserves, predict the most likely outcome to be even less calamitous: from RCP 2.6 to RCP 4.5 (Fig. 33.1). (Doose 2004; Kharecha and Hansen 2008; Brecha 2008; Nel 2011; Chiari and Zecca 2011; Ward et al. 2011, 2012; Höök and Tang 2013; Mohr et al. 2015; Capellán-Pérez et al. 2016; Murray 2016; Wang et al. 2017).

Why? Because the IPCC scenarios do not account for the fact that the era of abundant and affordable fossil fuels is drawing to a close. The IPCC RCP 8.5 hot-house world scenario assumes a fivefold increase in coal use by 2100 (Ritchie and Dowlatabadi 2017), even though coal production may have peaked, or will soon (see Chap. 6).

Good news, we are not cooked! Or at least we will be medium rare rather than well done.

If oil did peak in 2018, then the path may be closer to the IPCC RCP 2.6 scenario than RCP 4.5. The RCP 2.6 model requires carbon dioxide emissions to start to decline by 2020 and reach zero by 2100. RCP 4.5 posits CO_2 decline starting about 2045 and reaching half that level by 2100.

Of course, climate change is locked in for many centuries, with sea level rising, droughts, lowered food production, potential tipping points for the Amazon, and more. The IPCC reported that "about half of a CO_2 pulse to the atmosphere is removed over a time scale of 30 years; another 30% is removed within a few centuries; and the remaining 20% will typically stay in the atmosphere for many thousands of years" (Solomon et al. 2007).

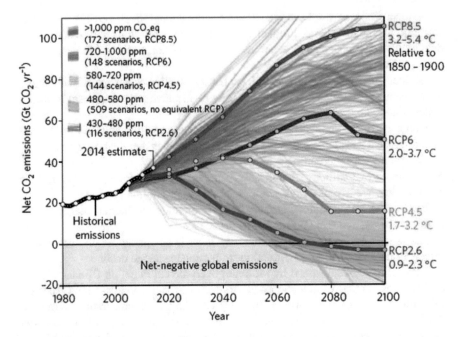

Fig. 33.1 IPCC Representative Concentration Pathways (RCPs) scenarios to 2100. Image credit: Neil Craik, University of Waterloo

Eventually, the earth will return to lower greenhouse levels as the ocean and land absorb carbon dioxide. Carbon sequestration, wind, solar, geo-engineering, and other remedies are trivial compared to the effect declining fossil fuels will have on reducing greenhouse gas emissions.

Climate change is also a symptom of overpopulation and overshoot of the planet's carrying capacity. If family planning became the green new deal, there would be a chance for all problems to be reduced in severity.

By burning fossil fuels, we are rapidly changing the atmosphere, the oceans, and the climate, triggering the potential extinction of millions of species. Our exponential growth and takeover of land and resources is also leading to a "sixth extinction." One way or another, declining fossils will mean a much lower human population. As human population levels decline, this will slow down the sixth extinction, climate change, freshwater depletion, decrease pollutants, and topsoil erosion. Too, there will be an end to the fishing out of our oceans. The days of diesel factory ships going after every last school of fish on earth with helicopters, airplanes, satellites, and sonar using large machines to spew out miles of nets and long lines to reel in tons of fish at a time are coming to an end. Can you think of a single problem that will not be better when there are fewer of us?

I owe you some more good news. Declining oil means you can stop worrying about robots taking over. What energy could they be built with and run on after fossils? Not that a robot overthrow was ever an issue. The human cortex is 600 billion times more complicated than any artificial network. The code to simulate the human brain would require hundreds of trillions of lines of code inevitably riddled with trillions of errors (Kasan 2011).

As a special bonus, I will ease your fears about a space alien invasion. As Sir Fred Hoyle (1964) wrote "With coal gone, oil gone, high-grade metallic ores gone, no species however competent can make the long climb from primitive conditions to high-level technology. This is a one-shot affair. If we fail, this planetary system fails so far as intelligence is concerned. The same will be true of other planetary systems. On each of them there will be one chance, and one chance only."

One way to cope with the grief of the party ending is to be thankful for what you have. It will be rough for all of us who have known the brief century of oil. But future generations will have never known anything else and they will be fine. Kansas pioneers survived and thrived despite locusts, floods, droughts, illness, and more (Stratton 1982). There are thousands of books about survival in hard times. History offers many lessons about how we can reinvent our way of life and find joy and meaning in simpler lives.

The Great Simplification—How to Make the Transition

The best way to manage our energy decline is to accept its inevitability and to embrace the challenge of transitioning to a simpler world.

It is a shame climate change gets all the attention given the many other existential threats (Rockström et al. 2009; Ripple et al. 2017). In the past 5 years, the New York Times has mentioned climate change over 15,000 times and "peak oil" just nine times. That upsets me! And while I have got a bee in my bonnet, the mainstream media also has given little coverage to existential issues such as soil erosion, soil degradation, groundwater depletion, deforestation, invasive species, biodiversity, nuclear war, ozone depletion, fishery collapse, eutrophication, overpopulation, ocean acidification, nuclear waste, phosphate depletion, carrying capacity, land-use change, overshoot, and decaying infrastructure. Okay, I am a glutton for punishment!

Fortunately, there are organizations with plans for making the transition as smooth as possible to a low energy world. Check out the websites of transitionnetwork.org, transitionus.org, postcarbon.org, resilience.org, energyandourfuture.org, and simplicityinstitute.org. Check out my website, energyskeptic.com and the categories "What To Do" and "Books." Also, see the last four pages of the German military peak oil study referenced in this footnote (BTC 2010). Want to find the best place to live given climate change and energy decline? Read Day & Hall's (2016) "America's Most Sustainable Cities and Regions."

If you are interested in becoming a prepper, you are in good company: Search the term prepping online and you will find 62 million results.

Covid-19 has taught us that food shortages and supply chain disruptions are not far-fetched. In 2020, even toilet paper was a precious commodity for a while. To cope with future energy and food shortages, now is a good time to prepare a rationing plan for the distribution of essential goods to everyone rather than just those who can afford high prices for food and gasoline. We should plan for stockpiles of food at home, town, and state levels. Thousands of small granaries should be built nationwide (DOE 1979; Golob et al. 2002; Cox 2013).

Transportation oil can be reduced by encouraging local and decentralized economies.

Odum and Odum (2001) have written perhaps one of the best books on coping with energy descent. Just a few of their ideas include: How to reorganize cities, limit luxury and waste, fewer private autos and more bicycles, people moving closer to work, a global effort to reforest depleted land in undeveloped countries, an income limit $150,000 a year rather than reducing the number of employees, more agricultural labor provided by people leaving the cities as employment decreases, ways to make the likely crash of the stock market less damaging when the public realizes growth is over, replacing flimsy housing with longer lasting structures, everything must be recyclable, and birth control. There, that should keep you busy!

Material scientists should be thinking about how to preserve knowledge if the electric grid and the ability to manufacture microchips can not continue. Without those microchips, we lose the Internet with its warehouses of computer servers and the cloud where we now store everything digital. Paper and microfiche last only 500 years when stored in optimal conditions, but they may be all we have in the future. Not to mention all our other dependencies on electricity.

Actions to prepare for energy descent would also dial down climate change. There are things we can do. Many things. Why not pay for the transition by cutting the military budget? We could start by getting rid of most of the 800 bases the US has overseas in more than 70 countries that cost over $100 billion a year (Vine 2015). US military expenditures are roughly the size of the next seven largest military budgets around the world, combined. Aircraft, tanks, ships, and weapons all require oil. It would be a shame to spend the remaining petroleum on wars rather than a transition.

Certainly, there is a chance of war over the remaining resources. Above all, nuclear weapons need to be scrapped and nuclear wastes buried (Rosenbaum 2011; Perry and Collina 2020). It is the least we could do for the grandchildren, who nobody speaks for.

Rather than releasing the four horsemen of the apocalypse, why not use the technology of family planning to get back in balance with the ecosystem? This would make birth control and abortion free and easy to get, encourage fewer children with tax incentives for one or no children, lower immigration in the US from one million people a year to far less to prepare for the lower carrying capacity of Wood World, and end unconditional birthright citizenship.

Make the maximum speed limit 55 mph, reduce consumption of everything, tax or ration gasoline, develop rationing plans for food, medicine, and other goods to use when it becomes necessary, subsidize high mileage cars and tax gas-guzzlers, insulate homes and buildings, create more bike lanes, unpave little-used roads (Fay and Kroon 2016), create canals for water transportation, build infrastructure that can last millennia like Roman roads and aqueducts since current roads have a lifespan of only 20 years, breed horses and oxen, and revamp schools to teach skills useful in Wood World like gardening. Use recycled materials!

Covid-19 has shown that people working from home rather than commuting to offices can dramatically lower energy use. As energy declines, cities with the best mass transit and walk or bicycle commuting will be in a better position to cope (Karlenzig 2008). Cities need to be redesigned to be less dependent on cars and have plans to reduce oil consumption by as much as the depletion rate to stay under the curve (Heinberg 2006).

It is helpful to have a goal to strive for. Transportation reform is a good place to start. For example, in my home town of Oakland, transportation is responsible for nearly half of greenhouse gas emissions and energy use. Cars can be shared, carpooling incentivized. Cities and towns also need to be rezoned so that farmers markets, local farms, grocery stores, drugstores, post offices, and other services are within walking or bicycling distance (OIAC 2008).

While we are at it, let us build wetlands to protect against sea-level rise, stormwater mitigation, and improve wildlife habitat. Bringing back the natural world will be a powerful antidote to our sense of loss.

Now that we know what lies in the future, our top priority should be the conservation of farmland. It should be a national policy to protect prime farmland from development. The best organic farms already know how to prevent soil erosion and to control pests without pesticides by using cover crops, mulching, and no-till, adding compost and manure, rotating and growing a wide variety of crops, and using biological pest control. Organic farming will also reduce our dependence on artificial nitrogen fertilizers, which are alarmingly increasing levels of nitrous oxide, a greenhouse gas 300 times more potent than CO_2 and lasting for over 100 years (Tian et al. 2020). The ingenuity these organic farms deploy is a wonder to behold. Organic farming should be adopted on industrial farms as quickly as possible.

Organic farming? Do I look like Pollyanna to you? Is not organic farming a flower power delusion, a pipedream never capable of being scaled up to feed the masses? Debates have raged for decades over whether organic agriculture has higher or lower crop yields than industrial agriculture. A major study has found that in most cases, diversified organic agriculture increases crop yields more food than monocrops. Tamburini et al. (2020) did a meta-analysis of 5188 studies with 41,946 comparisons between diversified and monoculture agriculture. The researchers found that diversification increased crop yields as well as biodiversity, pollination, pest control, nutrient cycling, soil fertility, carbon sequestration, and water regulation. Crop diversity reduced dependency on artificial nitrogen fertilizers. Diversification and organic farming also reduced soil and freshwater degradation, pollution, and greenhouse gases. Diversity was achieved with the rotation of numerous crops, hedgerows, seminatural habitats, reduced tillage, organic material soil amendments, and the inoculation of beneficial microorganisms into the soil.

Local agriculture and urban farms should be encouraged, with vacant properties converted to community gardens. Towns can, and already are, planting edible landscapes on public land, such as trees that bear fruit and nuts. Go for it: Rip up your lawn and plant a victory garden. Green waste is being collected in hundreds of cities to convert to compost to replace fertilizer. Some cities are already starting large food storage supplies. And as we know here in the "shake and bake" state of California, citizens should stockpile at least 3 weeks of food to cope with earthquakes, fires, power outages, energy shortage, and future climate change. Those of you from hurricane country also are familiar with the drill.

Given that we are going back to Wood World, we should stop exporting wood from our forests and instead get busy planting trees and regenerating our forests. That should be a top priority. At the rate we are currently destroying and degrading the world's forests, they will be gone within 100–200 years. The current rate of forest destruction gives us a 90% chance of an "irreversible collapse" of human civilization within 20–40 years (Bologna and Aquino 2020).

We will need to invent a new economic and political system to deal with a shrinking economy rather than the constant growth we depended on for two centuries as energy supplies kept growing. If we do not plan ahead, the default system is likely a mafia totalitarian state like Russia (Belton 2020; Gessen 2017). Call me a pessimist—okay, I had that coming—the US appears to be headed that way (Dean 2006; Ferguson 2012; MacLean 2017; Levitsky and Ziblatt 2018).

Really, we are collectively in denial. There is no preparation underway for energy decline at any level of government. As is evident with the US response to COVID-19, the effort really needs to be led, funded, and coordinated at the federal level. State and local preparation would help immensely, too. But since all levels of government struggle just to keep up with the day to day responsibilities, they seldom plan for the future.

Meanwhile, each of us needs to start preparing for a low energy world. There is good work to do to help our families and our community make this transition to the other side of the energy curve.

Clearly, our future geodestiny depends on our freshwater, forests, and topsoil. We should plan accordingly.

You knew that in the end I would bring up the fall of the Roman Empire. It is inevitable. We realists, the so-called pessimists and doomsters, always revert to the fall of the Roman empire. Such restraint—it is the first I have mentioned it. So, brace yourself: In "The History of the Decline and Fall of the Roman Empire," historian and author Edward Gibbon described the fall as due to a bloated and overextended military, widespread economic and political corruption, public apathy, and hedonism. And addiction to dependence on foreign resources.

Sound familiar? Not to worry, I know just the remedy. I heard it first from Wes "Scoop" Nisker (2007). He suggests the United States should just go to the United Nations and resign as a superpower and become an ordinary nation. We could relax, we would not have to work so hard. Look how happy Italians are today. We could launch a project called "The Great Leap Backward." And reopen the country as one vast theme park called "Formerly Great America."

References

Belton C (2020) Putin's people: how the KGB took back Russia and then took on the West. Farrar, Straus and Giroux

Bologna M, Aquino G (2020) Deforestation and world population sustainability: a quantitative analysis. Sci Rep 10:7631

BP (2020) Statistical review of world energy 2020, 69th edn. British Petroleum

Brecha RJ (2008) Emission scenarios in the face of fossil-fuel peaking. Energy Policy 36:3492–3504

BTC (2010) Armed Forces, Capabilities and technologies in the 21st century. Environmental dimensions of security. Sub-study 1. Peak oil security policy implications of scarce resources. Bundeswehr Transformation Centre, Future Analysis Branch. http://www.permaculture.org.au/files/Peak%20Oil_Study%20EN.pdf. Accessed 15 Nov 2020

Capellán-Pérez I, Arto I, Polanco-Martínez JM et al (2016) Likelihood of climate change pathways under uncertainty on fossil fuel resources availability. Energy Environ Sci 9:2482–2496

Chiari L, Zecca A (2011) Constraints of fossil fuels depletion on global warming projections. Energy Policy 39:5026–5034

Cox S (2013) Any way you slice it: the past, present, and future of rationing. The New Press

Dean JW (2006) Conservatives without conscience. Penguin Books

DOE (1979) Standby gasoline rationing plan. U.S. Department of Energy. https://doi.org/10.2172/6145884. Accessed 15 Nov 2020

Doose PR (2004) Projections of fossil fuel use and future atmospheric CO_2 concentrations, vol 9. The Geochemical Society Special Publications, pp 187–195

Fay L, Kroon A (2016) Converting paved roads to unpaved roads. Transportation Research Board of the National Academies of Sciences, Washington, DC

Ferguson CH (2012) Predator nation: corporate criminals, political corruption, and the hijacking of America. Crown Business

Gessen M (2017) The future is history: how totalitarianism reclaimed Russia. Riverhead Books

Golob P, Farrell G, Orchard JE (2002) Crop post-harvest: science and technology. Volume 1 principles and practice. Blackwell Science Ltd.

Hausfather Z, Peters GP (2020) Emissions – the 'business as usual' story is misleading. Nature 577:618–620

Heinberg R (2006) The oil depletion protocol: a plan to avert oil wars, Terrorism and Economic Collapse. New Society Publishers

Höök M, Tang X (2013) Depletion of fossil fuels and anthropogenic climate change – a review. Energy Policy 52:797–809

Hoyle F (1964) Of men and galaxies. Prometheus Books

Karlenzig W. 2008. Major US city post-oil preparedness for an oil crisis. Common Current. http://www.commoncurrent.com/pubs/MajorUSCityPost-OilPreparednessRanking.pdf. Accessed 15 Nov 2020

Kasan P (2011) A.I. Gone awry: the future quest for artificial intelligence. Skeptic. https://www.skeptic.com/reading_room/artificial-intelligence-gone-awry/. Accessed 15 Nov 2020

Kharecha PA, Hansen JE (2008) Implications of "peak oil" for atmospheric CO2 and climate. Glob Biogeochem Cycles 22:3

Levitsky S, Ziblatt D (2018) How democracies die. Crown

MacLean N (2017) Democracy in chains: the deep history of the radical right's stealth plan for America. Penguin Books

MEC (2020) Mining and mineral statistics: mineral usage statics. Minerals Education Coalition. https://mineralseducationcoalition.org/mining-mineral-statistics. Accessed 15 Nov 2020

Mohr SH, Wang J, Ellem G et al (2015) Projection of world fossil fuels by country. Fuel 141:120–135

Murray JW (2016) Limitations of oil production to the IPCC scenarios: the new realities of US and global oil production. Biophys Econ Resource Qual 1:13

NASA (2012) NASA-NOAA Satellite Reveals New Views of Earth at Night. National Aeronautics and Space Administration. https://www.nasa.gov/mission_pages/NPP/news/earth-atnight.html Accessed 10 Mar 2021

Nel WP (2011) A parameterised carbon feedback model for the calculation of global warming from attainable fossil fuel emissions. Energy Environ 22:859–876

Nisker W (2007) The decline and slide of the American empire. Freight & Salvage. https://www.facebook.com/watch/?v=343620603546928. Accessed 15 Nov 2020

Odum HT, Odum EC (2001) A prosperous way down: principles and policies. University Press of Colorado

OIAC (2008) Oil independent Oakland action plan. City of Oakland. http://library.uniteddiversity.coop/Energy/Peak_Oil/OIO-ActionPlan-020608.pdf. Accessed 15 Nov 2020

Perry WJ, Collina TZ (2020) The button: the new nuclear arms race and presidential power from Truman to trump. BenBella Books

Reuters (2019) U.S. dependence on China's rare earth: trade war vulnerability. https://www.reuters.com/article/us-usa-trade-china-rareearth-explainer/u-s-dependence-on-chinas-rare-earth-trade-war-vulnerability-idUSKCN1TS3AQ. Accessed 15 Nov 2020

Ripple WJ, Wolf C, Newsome TM et al (2017) World scientists' warning to humanity: a second notice. Bioscience 67:1026–1028

Ritchie J, Dowlatabadi H (2017) The 1000 GtC coal question. Are cases of high future coal combustion plausible? Energy Econ 65:16–31

Rockström J, Steffen W, Noone K et al (2009) Planetary boundaries: exploring the safe operating space for humanity. Ecol Soc 14:32

Rosenbaum R (2011) How the end begins: the road to a nuclear World War III. Simon & Schuster

Smil V (2010) Energy myths and realities: bringing science to the energy policy debate. AEI Press

Smil V (2017) Energy and civilization a history. The MIT Press

Solomon S, Qin D, Manning M, et al (2007) Technical summary. In: Climate Change 2007: The Physical Science Basis. Contribution of Working Group I to the 4th Assessment Report of the Intergovernmental Panel on Climate Change. Cambridge University Press

Stratton JL (1982) Pioneer women: voices from the Kansas frontier. Touchstone

Tamburini G, Bommarco R, Wanger TC et al (2020) Agricultural diversification promotes multiple ecosystem services without compromising yield. Sci Adv. https://doi.org/10.1126/sciadv.aba1715

Tian H, Xu R, Yao Y (2020) A comprehensive quantification of global nitrous oxide sources and sinks. Nature 586:248–256

USGS (2020) Mineral commodity summaries 2020. United States Geological Survey

Vine D (2015) Base nation: How U.S. Military Bases Abroad Harm America and the World. Metropolitan Books

Wang J, Feng L, Tang X et al (2017) Implications of fossil fuel supply constraints on climate change projections: a supply-side analysis. Futures 86:58–72

Ward JD, Werner AD, Nel WP et al (2011) The influence of constrained fossil fuel emissions scenarios on climate and water resource projections. Hydrol Earth Syst Sci 15:1879–1893

Youngquist W (1997) Geodestinies: the inevitable control of earth resources over nations and individuals. Natl Book Co.

Index

Printed in the United States
by Baker & Taylor Publisher Services